Pursuit of Power

NASA's Propulsion Systems Laboratory No. 1 and 2

By Robert S. Arrighi

National Aeronautics and Space Administration

NASA History Program Office
Public Outreach Division
NASA Headquarters
Washington, DC 20546

SP–2012–4548
2012

Table of Contents

Introduction vii
    Condensed History of the PSL ix
    Preserving the PSL Legacy ix
    Endnotes for Introduction xi

Acknowledgments xiii

Chapter 1: Legacy of Aircraft Engine Research 3
    Flying High with Feet on the Ground 3
    A Call for Nationalization 4
    Low-Pressure System 5
    Producing Results 6
    Barometer of the Future 6
    Engine Studies in Cleveland 7
    Pulling Ahead 9
    Endnotes for Chapter 1 11

Chapter 2: The Next Big Thing 13
    NACA's Panel of Experts 14
    Congressional Authorization 15
    The Designers 16
    Eugene Wasielewski 17
    Propulsion Systems Laboratory Rising 19
    Executing the Plans 28
    Assembling the Crew 28
    Endnotes for Chapter 2 30

Chapter 3: Harnessing PSL's Muscle 33
    Making Research a Reality 34
    Conducting a Test at the Propulsion Systems Laboratory 37
    Endnotes for Chapter 3 51

Chapter 4: Cold War Weapons 53
    The Nuclear Navaho 54
    Navaho's 48-Inch-Diameter Powerplants 54
    Bob Walker 56
    Getting the Damn Engine to Ignite 57
    Taking Advantage of the Opportunity 61
    Self-Immolation 63
    Coda 63
    Defense Missiles 64
    Howard Wine 66
    Testing the Turbojets 67

    General Electric 1950s Successes 67
    Ill-Fated Avro Project 68
    Endnotes for Chapter 4 70

Chapter 5: The Rocket Era 75
    Cleveland Rockets 75
    Transformation 76
    Hypersonic Heating 77
    John Kobak 80
    Neal Wingenfeld 81
    Homemade Rockets 82
    Star of the Rocket Engine Era 85
    The RL-10 Gets Worked Over 87
    Sitting on Top of All That Hydrogen 88
    Rocket Division Arrives 91
    Primitive Propulsion 91
    End of an Era 94
    Endnotes for Chapter 5 95

Chapter 6: Jet Engines Roar Back 99
    Airbreathing Research Personnel 100
    Supersonic Calibration 103
    Artificial Distortion 105
    Treatments 106
    Slower, Better, Cheaper 107
    The Compass Cope 108
    Hazards Exposed 110
    Endnotes for Chapter 6 112

Chapter 7: The Third Step 117
    Another Giant Emerges 117
    Flutter 124
    Full-Scale Engine Programs 125
    Pratt & Whitney F100 125
    J85-21 126
    Highly Maneuverable Aircraft Technology 127
    Fly by Wire 128
    Remote Control 129
    Multivariable Control 130
    Complete Control 131
    Glory Days 132
    Endnotes for Chapter 7 133

Chapter 8: No Tomorrow 137
    The Long Winter 137
    Demolition Decision 138

    Pulling PSL Down 140
    Historical Mitigation 145
    PSL Legacy...147
    Endnotes for Chapter 8 148

Bibliographic Essay..149

Interview List 151

Website 153

NASA History Series 155
    Reference Works, NASA SP–4000 Series..155
    Management Histories, NASA SP–4100 Series 156
    Project Histories, NASA SP–4200 Series...157
    Center Histories, NASA SP–4300 Series 158
    General Histories, NASA SP–4400 Series 159
    Monographs in Aerospace History, SP–4500 Series 160
    Electronic Media, SP–4600 Series..162
    Conference Proceedings, SP–4700 Series 163
    Societal Impact, SP–4800 Series..163

List of Images 165

Index..169

Introduction

As the Sun set on a pleasant late September evening in 1952, the exterior lights of the new Propulsion Systems Laboratory (PSL) were illuminated. A photographer and his assistant set up their tripod and camera at several locations around the facility, attempting to capture the sprawling tangle of steel, pipes, and valves. By the time that they had made it across the street to snap the two wide-angle shots, darkness had fallen and the most modern engine testing facility in the country shone brightly against the night sky over the National Advisory Committee for Aeronautics (NACA) Lewis Flight Propulsion Laboratory in Cleveland, Ohio.

There were others gazing into the Midwestern darkness that evening. The Cold War had firmly taken root, and world tensions were elevated. Local teams of "skywatchers" from the Ground Observation Corps were scanning the night sky for enemy aircraft. The range and destructiveness of weaponry had increased dramatically in the 11 years since Pearl Harbor. "For the first time in our history," one NACA Lewis skywatcher declared, "a potential enemy has the power to make sudden, devastating attacks on any part of our country."[1]

In just weeks, United Nations forces would launch the ill-fated Battle of Triangle Hill—the last major offensive of the Korean War; General Dwight Eisenhower would be elected President, and the first hydrogen bomb would be detonated in the Marshall Islands.

Aerospace development also was advancing dramatically. Jet engines were demonstrating tremendous increases in power, U.S. and Soviet engineers were racing to develop intercontinental ballistic missiles, and Harvey van Allen was beginning to circulate his blunt-body theory for atmospheric reentry. In addition, the International Geophysical Year had been set for 1957, and *Colliers* had captivated the public's imagination with a cover story titled "Man on the Moon."[2]

In Cleveland, construction of the PSL, the nation's largest facility for testing full-scale engines in simulated flight conditions, was completed. As the photographers documented the shining exterior on that September evening, a ramjet for the Navaho Missile program and a General Electric J73 turbojet for the F86H Sabre fighter were being installed inside its two test chambers.

Image 1: PSL No. 1 and 2 on the evening of 24 September 1952. (NASA C–1952–30764)

There were several ways to test engines, but test stands provided the most efficient and useful method. A single stand could be used to study many types of engines in an environment that was safer and more efficient than that on research aircraft. Efficiency improved because installing instrumentation and measuring thrust is more difficult in flight research.[3]

Pilots found early on that engines behave differently in altitude conditions than at sea level, and engineers quickly determined that the test stands needed to simulate altitude conditions in order to properly test engines. In general, this meant reducing the temperature and pressure of the air and creating velocity. Removing moisture from the air and expelling the engine's exhaust were also important.

Unlike wind tunnels, these altitude chambers were generally used to study only the engines, not the engine cowlings or mounts. Another difference was that the amount of power and air required was far less than for tunnels because the air was ducted only through the engine—not through the surrounding test section.

PSL's two chambers, referred to as "PSL No. 1 and 2," could simulate the internal airflow conditions experienced by the nation's most powerful engines over a full range of power and altitude levels. This allowed researchers to analyze the engine's thrust, fuel consumption, airflow limits, combustion blowout levels, acceleration, starting characteristics, and an array of other parameters. The range of PSL's studies was later expanded to include noise reduction, flutter, inlet distortions, and engine controls.

There were three distinct eras during PSL No. 1 and 2's operating years, each with its own group of researchers: NACA Lewis's Engine Research Division managed the ramjet and turbojet period of the 1950s, the Chemical Rocket Division conducted most of PSL's research in the 1960s, and the Airbreathing Engines Division assumed control for the turbofan and supersonic inlet studies of the late 1960s and 1970s. The researchers and test engineers in these divisions developed and ran the tests. The mechanics and technicians in the Test Installations Division maintained the PSL and integrated the test equipment into the facility throughout the years.

Image 2: The PSL No. 1 and 2 facility viewed from the west on 9 June 1959. The Shop and Access Building containing the two test chambers is to the left, and the larger Equipment Building containing the exhausters and compressors is behind and to the right. (NASA C–1959–50861)

Condensed History of the PSL

NACA management initiated its plans for the PSL in November 1947, and construction started two years later. When the facility began operating in October 1952, PSL No. 1 was used for turbojet studies while PSL No. 2 concentrated on ramjets.

By the late 1950s, Pratt & Whitney, Wright Aeronautical, and the U.S. Air Force began building their own propulsion labs and altitude facilities. The PSL remained a vital resource by continually upgrading its two chambers, control room, and air-handling system. The installation of a pebble bed heater in the late 1950s permitted hypersonic studies, and the installation of a flamespreader in the mid-1960s allowed more powerful engines to be tested without damaging the cooling equipment.

By the 1960s rocket systems of increasing complexity were being studied in both chambers, including an extensive investigation of the Pratt & Whitney RL-10 in PSL No. 1. In the late 1960s, the PSL again turned to airbreathing engines for aircraft. Unlike the studies in the 1950s, this new effort included propulsion systems for civilian aircraft.

In 1967, construction was undertaken on a new PSL building with two additional, more powerful altitude chambers, referred to as "PSL No. 3 and 4." All four PSL chambers were used for turbojet and turbofan studies from 1972 to 1979. Budget concerns led to the ultimate shuttering of PSL No. 1 and 2 in 1979.

After years of idleness that included the installation of temporary office space around the test chambers, the NASA Glenn Research Center[a] decided to remove the original facility in 2004. Five years later, the PSL No. 1 and 2 facility was demolished. The PSL No. 3 and 4 facility and the Equipment Building, since renamed the Central Air and Equipment Building, continue to operate today. The PSL remains the National Aeronautics and Space Administration's (NASA's) sole facility for testing full-scale aircraft engines in simulated flight conditions.

The PSL proved to be a robust test facility that could keep pace with the relentless advance of aerospace technology over the decades. The original chambers were versatile enough to study emerging propulsion systems such as the turbojet, ramjet, chemical rocket, and turbofan engines. Work in the PSL on the RL-10 rocket engine was essential to the success of the Centaur Program. The PSL served as a key component in the center's 65-year history of testing engines at altitude conditions.

Preserving the PSL Legacy

The NASA Glenn History Office has undertaken the task of documenting many of its historic facilities. Histories of the Icing Research Tunnel, Plum Brook Reactor Facility, Rocket Engine Test Facility, and Altitude Wind Tunnel have been published. These books demonstrate the significance of each facility to the nation's aerospace community while sharing personal stories of some of the unsung researchers, mechanics, and technicians who performed groundbreaking research and made the giant facilities run. It is hoped that this publication continues this tradition.

The PSL No. 1 and 2 facility was determined to be eligible for, but was not listed on, the National Register for Historic Places. Glenn's History Program and Historic Preservation Officer worked with the Ohio State Historic Preservation Officer to develop a plan to document PSL's contributions and distribute that information to the public.[4] This effort included collecting documents from Glenn Records Management holdings, the History Office archives, retirees, and other sources. Hundreds of photographs, films, and documents were digitized. In addition, a thorough photographic survey was performed, and two graphical renderings of the facility were created.

The collected information was distilled for the public and NASA communities into this publication, an exhibit display, and a website to be shared with the

---

[a] The Cleveland laboratory began operation in 1942 as the NACA Aircraft Engine Research Laboratory (AERL). In 1947 it was renamed the NACA Flight Propulsion Laboratory to reflect the expansion of the research. In September 1948, following the death of the NACA's Director of Aeronautics, George Lewis, the name was changed to the NACA Lewis Flight Propulsion Laboratory. On 1 October 1958, the lab was incorporated into the new NASA space agency, and it was renamed the NASA Lewis Research Center. Following John Glenn's flight on the space shuttle, the center name was changed again on 1 March 1999 to the NASA Glenn Research Center.

Pursuit of Power

Image 3: This 7- by 10-foot exhibit display created in 2009 describes PSL No. 1 and 2 and some of the tests conducted there. (NASA P–0931 PSL 1 2)

public and NASA employees. For more detailed information, research materials, and hundreds of additional photographs and videos visit http://pslhistory.grc.nasa.gov.

Endnotes for Introduction
1. "Skywatchers Needed," *Wing Tips* (19 September 1952).
2. "Man on the Moon," *Collier's Weekly* (18 October 1952).
3. "Propulsion Systems Laboratory Test Chambers #3 & #4 Facility Description," June 1982, NASA Glenn History Collection, Test Facilities Collection, Cleveland, OH.
4. Leslie Main, Dallas Lauderdale, and Mark Epstein, Memorandum of Agreement between the National Aeronautics and Space Administration and the Ohio State Historic Preservation Office regarding the Demolition of Propulsion Systems Laboratory, Cell No. 1 and Cell No. 2, 4 September 2007, NASA Glenn History Collection, Oral History Collection, Cleveland, OH.

· Acknowledgments ·

This publication completes the 2005 agreement between NASA Glenn's Logistics and Technical Information Division and the Facilities Division to document the Propulsion Systems Laboratory No. 1 and 2 and the Altitude Wind Tunnel prior to their demolition. Former Glenn History Officer Kevin Coleman and Historic Preservation Officer Joe Morris were instrumental in forging this agreement with the assistance of former contract manager Ralph Bevilacqua.

Anne Mills and Les Main have long since stepped into the respective positions of History Officer and Historic Preservation Officer and have been the guiding forces behind this publication. Anne's enthusiastic support for not only this project but all of the Glenn History Office activities is greatly appreciated. Anne also graciously serves as the History Office's public face so that I can concentrate on research and the archives. Les has generously supported a number of historical preservation projects and has provided insight and assistance with numerous nonmitigation issues. I am grateful for Les's guidance over the years and look forward to continued collaboration with him in the future.

This publication would not have been possible without the assistance of fellow Wyle Information Services employees on the Technical, Information, Administrative and Logistics Services (TIALS) contract. Their efforts are deeply appreciated. Nancy O'Bryan, Kelly Shankland, and Lorraine Feher once again provided excellent copy editing, design/layout, and proofreading for the book. Debbie Demaline helped locate documents in the Records Management System, and Suzanne Kelly returned hundreds of negatives to storage. Mark Grills and Michelle Murphy scanned these negatives, while Bridget Caswell and Quentin Schwinn photographed the facility prior to and during its demolition. Don Reams and Caroline Rist provided oversight for the project, particularly in its critical final stages.

I thank Virginia Dawson and Mark Bowles, whose previous research contributed significant scholarly insights and source material regarding the history of the center. I also thank former Garrett Corporation engineers Jerry Steele and John Evans for taking such a keen interest in my out-of-the blue inquiry regarding the ATF3 tests.

Thanks go to Tina Norwood and the anonymous peer reviewers who took the time to thoroughly review the manuscript and make constructive suggestions that elevated the quality. The efforts of Steve Garber as the liaison for the reviews and coordinator for the printing are appreciated, as are the efforts of Bill Barry and Julie Ta, all from the NASA History Program Office.

Although it has been several years since I have spoken to some of the retired NASA employees noted here, I remain grateful for their time and generosity. Thanks go to Bob Walker, Howard Wine, Bill Harrison, Frank Holt, and John Kobak. I particularly thank Neal Wingenfeld for not only discussing his experiences with me several times, but taking time to thoroughly review the manuscript.

Most importantly, I thank my wife Sarah and my family Bob Sr., Gerri, Michele, Dan, Autumn, and Kim for their continued love and support throughout the years.

Image 4: An engine is tested in the Engine Propeller Research Building ("Prop House") on 15 April 1944. The Prop House was the first in a long line of engine test facilities at the NACA Aircraft Engine Research Laboratory (AERL) in Cleveland, Ohio. (NASA C–1944–04598)

· 1 ·
Legacy of Aircraft Engine Research

Lanky U.S. Army pilot Major Rudolph Schroeder guided his Lepere C-11 biplane ever higher over Southwestern Ohio on 27 February 1920. As the aircraft neared a record-breaking altitude of 31,000 feet, Schroeder began losing consciousness. The veteran pilot quickly removed his goggles to examine the malfunctioning oxygen regulator but was confronted with another peril. His eyeballs instantly froze in the −63°F air found at the high altitude. Blacking out and nearly blind, Schroeder put the aircraft into a 300-mile-per-hour, 6-mile dive. At the very last moment, Schroeder awoke and regained control. He somehow pulled the aircraft out of its plummet and blindly navigated to the McCook Field airstrip.[1]

Schroeder's near-calamity was not the first time that altitude had affected pilots or aircraft. Schroeder himself had noted previously that high-altitude flights left him feeling sleepy, tired, cross, and hungry.[2] Aircraft engines had a similar reaction in the frigid, oxygen-deprived atmosphere above 10,000 feet. Fuels coagulated, spark plugs developed carbon deposits, and combustion efficiency decreased. In general, the climb rate decreased as altitude increased until the aircraft could climb no more.[3]

Schroeder's breathtaking 1920 mission was also notable because it was the first flight of a turbo-supercharger. The innovative technology ushered in a new era of high-altitude flight. In 1917 the NACA asked General Electric to develop a device to enhance high-altitude flying. General Electric developed the turbosupercharger in response to a 1917 request from the NACA to develop a device to enhance high-altitude flying. The device pushed larger volumes of air into the engine manifold. The extra oxygen allowed the engine to operate at its optimal sea-level rating even when at high altitudes. Thus, the aircraft could maintain its climb rate, maneuverability, and speed as it rose higher into the sky. The technology proved itself on Schroeder's 1920 flight.[4]

Flying High with Feet on the Ground

Aeronautical engineers had been exploring methods of testing their engines in altitude conditions for several years. It appears that the first altitude testbeds were built by German engineers on mountain sides at elevations of 5,000 to 6,000 feet. Frequent inclement weather and the relatively low altitudes, however, tempered the usefulness of these facilities. The Italian manufacturer Fiat advanced the field by creating a sea-level engine testbed that could simulate altitudes up to 16,000 feet.[5]

By 1916 Germany's Zeppelin Aircraft Works had built an engine test facility that could simulate the low pressures of altitudes up to 20,000 feet. A valve-regulated blower evacuated air from the room, cooling water was carefully controlled, and the hot exhaust was removed.[6]

Image 5: The 6-foot, 3-inch "Shorty" Schroeder stands next to his Lepere C-11 biplane, the nation's first aircraft dedicated solely to research. Schroeder broke four altitude records between 1918 and 1920. The eye injury on the February 1920 flight ended his flying career. (U.S. Air Force Museum)

An NACA report noted that the altitude facility "is the result of long and costly experiments. Considering that it would scarcely be worthwhile for the majority of aviation engine manufacturers to construct altitude benches of their own, though future needs will make it urgently advisable to investigate engines with regard to altitude requirements, the Zeppelin Dirigible Works have offered their vacuum chamber and its measuring installation for rent for such purposes."[7]

A Call for Nationalization

Germany, France, Russia, and Britain had established publicly funded aeronautical research programs in the first decades of flight. The resulting cooperation between European governments and their research institutions helped the European aviation industry outpace the Americans. As early as 1911, some in the U.S. aviation field were clamoring for a coordinated research effort, but at the outset of World War I, the nation's aviation efforts remained unorganized.

On 3 March 1915 Congress decided to act. It was just nine weeks before the sinking of the *Lusitania*, and war clouds were gathering over the capitol. The NACA was established through a U.S. Navy appropriation to remedy the situation. The NACA's unpaid Executive Committee, consisting of military, industry, university, and other aviation experts, did not perform research itself but advised existing entities on future research trends.[8] One of its board members was William Stratton, head of the National Bureau of Standards (NBS).

Image 6: William Stratton, Chair of the NACA's Powerplants Committee from 1916 to 1931 at the first meeting of the NACA on 23 April 1915. (NASA A-6786; image was cropped to just show Stratton)

Image 7: The NBS campus in Washington, DC, just 3 miles northwest of the White House. The NBS, founded in 1901, tested virtually every imaginable type of technology. In the 1910s and 1920s it investigated almost all components of an aircraft, including powerplants. (National Institute of Standards and Technology)

The NBS served as the military's primary research laboratory at the time. A torrent of research requests flooded into the NBS following the nation's entry into the war in April 1917. No requests were more important than those regarding flight. The United States was lagging far behind both the Germans and the Allies in aeronautics. The small fleet of aircraft that the United States did possess relied largely on European technology.[9]

The NBS tested aircraft components that involved aerodynamics, materials, instrumentation, and most importantly, propulsion. Although nearly every element of the U.S. aircraft needed improvement, the engines and their behavior at altitude required particular attention.

In June 1917 a military board assigned industry engineers the task of quickly designing a new American aircraft engine, the Liberty. Just one month later the NBS commenced several months of testing of 8- and 12-cylinder Liberty engines. The Liberty-12 was the nation's first standardized and mass-produced aircraft engine, the first to use a turbosupercharger, and America's most important contribution to World War I aviation.[10]

Low-Pressure System

Special emergency wartime funding in April 1917 permitted the expansion of the staff and facilities at the NBS's Washington, DC, campus. The new facilities included a wind tunnel, a dynamometer, and an Altitude Laboratory for operating an aircraft engine in simulated altitude conditions.[11]

Stratton asked Hobart Cutler Dickinson, a refrigeration and insulation specialist, to lead the NBS aircraft engine studies and design the new altitude chamber.[12] The basic design of Dickinson's Altitude Laboratory, the nation's first, was similar to the Zeppelin facility. The temperatures and pressures within the chamber, however, could be altered to simulate an altitude of 35,000 feet.[13]

The 6- by 15-foot facility was designed to handle the largest engines currently available. The 1-foot-thick reinforced-concrete walls were lined in cork. Doors on either side allowed researchers to work on the engine between runs. They could view the engine during the tests through three windows in each door.

Image 8: A Liberty-12 engine inside the NBS Altitude Laboratory, which has one of its side doors open. The chamber was originally in a temporary stucco building. It was later moved and placed with a second chamber in a permanent brick structure named the Dynamotor Laboratory. (National Institute of Standards and Technology)

Pursuit of Power

The engine was mounted on a 300-horsepower dynamometer test stand to measure its thrust. A centrifugal blower removed the engine's exhaust and maintained the chamber's low atmospheric pressure. A refrigeration unit cooled the air entering the chamber, and fans provided circulation.[14]

Cables and wires connected the spark, throttle, and instrumentation devices to an external control panel. Oil, gasoline, cooling water, and carburetor air were fed to the engine. This basic concept would be expanded on at several NACA facilities, including the Propulsion Systems Laboratory (PSL).[15]

Producing Results

In December 1917 the NBS commenced a series of tests of the Liberty engine in its new Altitude Laboratory. The War Department assigned an NACA representative to oversee the project. The Liberty-8 engine gave out several weeks later after just seven problematic runs at a simulated 25,000 feet. It was replaced in late January 1918 by a Hispano-Suiza engine. In just one week, the Suiza was run almost 1,000 times at altitudes up to 30,000 feet. The NBS thus completed the nation's first comprehensive set of observations of a functioning aircraft engine in simulated altitude conditions.[16]

The Altitude Laboratory continued testing throughout the year on the Liberty-12 and additional Hispano-Suiza engines. The altitude investigations of the 1920s demonstrated the performance of the carburetor, compressors, radiator, fuels, oils, and superchargers at higher altitudes. It was also the first time that different types of gasoline were analyzed for their performance. The NBS studies provided basic understanding of the engine performance at high altitudes.[17-20]

The NACA established its own research laboratory at Langley, Virginia, in 1919; but the NBS continued to oversee almost all U.S. aircraft engine research. The NACA created a Powerplants Committee, chaired by Stratton from 1916 to 1931, to address propulsion concerns, but the NACA performed comparatively little engine research.

As aircraft engines grew more powerful, there was an increased need to study them in altitude conditions. The NBS's Altitude Laboratory yielded excellent results, but newer engines were exceeding its capabilities. The development of turbosuperchargers in the early 1920s only increased the need for better altitude test facilities. Again the Europeans led the way.

Barometer of the Future

In 1933 the Germans created a test facility at Berlin-Aldershof capable of partially simulating altitudes of 62,000 feet with independently controlled temperature and pressure. A second German facility at Rechlin could create a pressure altitude of 59,000 feet. A unique refrigeration method autonomously reduced the inlet temperature to −60°C.[21]

NACA Executive Committee members Charles Lindbergh and John Jay Ide resided in Europe during the interwar years and reported on Germany's technological advances. In the mid-1930s George Lewis, the NACA's Director of Aeronautical Research, toured German research laboratories several times. Lewis was amazed at the size and complexity of the research agencies, which, unlike the NACA, focused on aircraft engines as well as aerodynamics. He noted one facility that was able to test water and air-cooled engines under simulated altitude conditions. The intake air and exhaust were kept at pressure and temperature levels that corresponded to the altitude conditions.[22]

The German technological advances were coupled with political blustering and militarization. George Lewis noted, "Everything in Germany is considered first from the military point of view."[23] He and his fellow NACA Executive Committee members realized that the United States needed to quickly expand its own aeronautical research efforts. In response to the NACA's recommendation, Congress authorized the creation of a new NACA lab at Sunnyvale, California,[b] in 1939 to study aspects of high-speed flight. Almost immediately thereafter, it approved a third laboratory to concentrate entirely on aircraft engines and propulsion.

---

[b] It was called the NACA Ames Aeronautical Laboratory until 1958 and the NASA Ames Research Center thereafter.

Construction of this new Aircraft Engine Research Laboratory[c] (AERL) in Cleveland, Ohio, began in early 1941. Unbeknownst to the NACA, the Germans had flown the first jet aircraft in August 1939, and the nature of aviation would be forever changed before the AERL was complete.

Engine Studies in Cleveland

Although late to the field of propulsion, the NACA soon surpassed not only the NBS but many of the foreign research centers. The AERL emerged from its construction in 1944 as the nation's predominant aircraft engine testing laboratory. Its massive facilities dwarfed the NBS engine facilities. Former AERL thermodynamicist, Benjamin Pinkel, told historian Virginia Dawson in 1991, "We had one of the best [engine] laboratories in the world from the equipment standpoint…It was the first one in our country."[24]

The AERL's first major facility to come online was the Engine Propeller Research Building, or "Prop House," in May 1942. The Prop House contained four 24-foot-diameter test stands to test 4,000-horsepower piston engines in ambient conditions. Although the Prop House yielded some useful data, it was only a larger version of test stands located at other research laboratories. (For a complete history of the lab, see Dawson.[25])

Another facility, the Engine Research Building (ERB), contained over 60 smaller test cells and rigs for studying engine components and single cylinders. This type of research required only a limited amount of

---

[c]The Cleveland laboratory began operation in 1942 as the NACA Aircraft Engine Research Laboratory (AERL). In 1947 it was renamed the NACA Flight Propulsion Laboratory to reflect the expansion of the research. In September 1948, following the death of the NACA's Director of Aeronautics, George Lewis, the name was changed to the NACA Lewis Flight Propulsion Laboratory. On 1 October 1958, the lab was incorporated into the new NASA space agency, and it was renamed the NASA Lewis Research Center. Following John Glenn's flight on the space shuttle, the center name was changed again on 1 March 1999 to the NASA Glenn Research Center.

Image 9: Crew at work on an Allison V-1710 engine mounted on a dynamotor rig in the east wing of the ERB on 19 March 1943. The building had many test cells for studying engine components. (NASA C–1943–01345)

Image 10: An NACA-designed ramjet engine with afterburner is installed in the AWT on 7 February 1946. To produce the proper settings, researchers changed the design so that the conditioned air was ducted directly to the engine's inlet. (NASA C–1946–14244)

combustion airflow. Both the Prop House and ERB were markedly affected by the arrival of the new turbojet technology. Former AERL civil engineer Harold Friedman explained, "Jet engines didn't come in single cylinders. You had to test the whole engine. So they needed new air supply and exhaust supply."[26]

The AERL's next major engine testing facility, the Altitude Wind Tunnel (AWT), was truly revolutionary. The AWT took the concept of the altitude chamber and incorporated the benefits of a wind tunnel. The facility could operate full-scale engines in conditions that replicated the speed, altitude, and temperature of actual flight. The facility included a massive refrigeration system, large exhauster equipment, an air dryer, exhaust scoop, and a 20-foot-diameter test section. Although designed for piston engines, from its very first runs in February 1944, the AWT was adapted to accommodate turbojets. The key alteration was the ducting of the conditioned airflow directly to the engine inlet.[27]

It soon became apparent that altitude affected jet engines even more than piston engines. Pinkel explained, "We had to have a way to test these engines at high speeds and at high altitudes on the ground because it made a big difference in their performance, particularly with regard to the performance of the combustion chamber, because they had a tendency to blow out (flameout) at high altitudes. When they blew out, it was serious because restarting in the thin air of high altitudes could be difficult."[28]

The AWT was the most advanced engine testing facility in the nation and an indispensable tool for the new technology. The massive influx of requests to

Image 11: One of the two 10-foot-diameter altitude chambers in the Four Burner Area in the ERB on 31 January 1956. The facility served as the bridge between the AWT and PSL. (NASA C–1947–19780)

study different turbojet engines in the AWT resulted in an 8- to 12-month backlog following World War II. To alleviate this problem, a new facility, referred to as the "Four Burner Area," was constructed inside the ERB. The Four Burner Area contained two 10-foot-diameter test cells in which full-scale engines could be run at simulated altitudes of 50,000 feet.[29]

The facility engineers forewent the wind tunnel design in favor of test chambers. This meant that the researchers would not be able to study the engine's integration with the wing or nacelle. The chamber design was more efficient for engine testing, however. Pinkel explained, "A wind tunnel is awfully expensive and costly to run. The wind tunnels handled much more air than was needed for the engine. [In contrast,] the altitude chambers restricted their power expenditure to just the air needed by the engine being tested."[30]

Although only one cell could be run at a time, the two chambers in the Four Burner Area allowed two engines to be installed concurrently. When it became operational in 1947, the Four Burner Area not only eased the AWT's burden, but its compressors increased the capabilities of the AERL's overall central-air-handling system.[31]

Pulling Ahead

NACA researchers used the Four Burner Area altitude tanks extensively during the late 1940s and 1950s to study early generations of the turbojet. However, the size and capabilities of jet engines had increased dramatically since NACA engineers first began designing the facility in 1945.

The early turbojets handled only 25 pounds of air per second and produced 1,500 pounds of thrust. By the early 1950s engines were handling over 100 pounds of air per second and generating over 6,000 pounds of thrust. The new 10,000-pound-thrust engines were just around the corner.[32]

Almost immediately after the Four Burner Area chambers began operating in 1947, the NACA's Cleveland laboratory undertook an effort to upgrade its air-handling systems and construct a new and even larger altitude test facility to accommodate these ever-growing engines. The new altitude facility was originally referred to as the "Propulsion Science Laboratory" but would soon become the "Propulsion Systems Laboratory (PSL)."

Former PSL Crew Chief Bob Walker remembered, "We had the [Four Burner Area] test cells at Lewis that were pretty modern and they performed a lot of good work, but they were under capacity for the kind of jet engines and airflows that were required. So the PSL No. 1 and 2 was the big step at that particular time."[33]

In fact, PSL's strong capabilities eventually led to the repurposing of the AWT. Howard Wine, another former PSL crew chief, explained, "I think that at the time PSL promised the opportunity to get into various jet and rocket engines that AWT didn't have. It wasn't hooked up to the compressors and exhaust system to the degree that PSL was, and PSL could handle larger engines, larger volumes of flow than AWT."[34]

The PSL would be the preeminent facility for full-scale engine testing, not only at the NACA's Cleveland lab but in the entire nation. "We were supposed to be the propulsion guys," Walker explained. "We were building propulsion facilities, and [PSL] was definitely a propulsion facility."[35]

Image 12: General Electric J73 in PSL No. 1 on 16 October 1952. The 14-foot-diameter PSL chambers could test full-scale engines at simulated altitudes of up to 70,000 feet. (NASA C–1952–30959)

Endnotes for Chapter 1

1. Robert van Patten, "Pioneers at High Altitude," *Air Force Magazine*, April 1991.
2. van Patten, "Pioneers at High Altitude."
3. General Electric, "The Turbosupercharger and the Airplane Power Plant" (General Electric Technical Manual TM 1-404, 1943) formatted by Randy Wilson (1997), http://rwebs.net/avhistory/opsman/geturbo/geturbo.htm (accessed 15 May 2011).
4. General Electric, "The Turbosupercharger and the Airplane Power Plant."
5. Kyrill von Gersdorff, "Aeroengines—Altitude Test Beds," *Aeronautical Research in Germany: From Lilienthal Until Today*, 147, Ernst-Heinrich Hirschel, Horst Prem, and Gero Madelung, editors (Berlin: Springer-Verlag, 2004).
6. W.G. Noack, *Tests of the Daimler D-Iva Engine at a High Altitude Test Bench* (Washington, DC: NACA TN 15, 1920).
7. Noack, *Tests of the Daimler D-Iva Engine*.
8. Alex Roland, *Model Research*, Volumes 1 and 2 (Washington, DC: NASA SP–1403, 1985), chap. 1.
9. Rexmond Cochrane, *Measures for Progress: A History of the National Bureau of Standards* (Washington, DC: U.S. Department of Commerce, 1966).
10. Cochrane, *Measures for Progress*.
11. Cochrane, *Measures for Progress*.
12. Robert Zarr, "The Testing of Thermal Insulators," *A Century of Excellence in Measurements, Standards, and Technology, A Chronicle of Selected NBS/NIST Publications, 1901–2000*, David R. Lide, editor (Washington, DC: NIST SP–958, 2001), pp. 10–13.
13. Cochrane, *Measures for Progress*.
14. "Third Annual Report of the National Advisory Committee for Aeronautics, 1917" (Washington, DC: NACA AR–3, 1918).
15. Hobart C. Dickinson and H. G. Boutell, "The Altitude Laboratory for the Test of Aircraft Engines" (Washington, DC: NACA No. 44, 1920).
16. Dickinson and Boutell, "The Altitude Laboratory for the Test of Aircraft Engines."
17. Cochrane, *Measures for Progress*.
18. Hobart C. Dickinson, "Performance of Aeronautic Engines at High Altitudes," *Third Annual Report of the National Advisory Committee for Aeronautics, 1917* (Washington, DC: NACA AR-3 Report No. 23, 1918).
19. NACA, "Fourth Annual Report of the National Advisory Committee for Aeronautics, 1918" (Washington, DC: Government Printing Office).
20. Eugene M. Emme, *Aeronautics and Astronautics: An American Chronology of Science and Technology in the Exploration of Space, 1915–1960* (Washington, DC: NASA Headquarters, 1961), pp. 1–11.
21. Kyrill von Gersdorff, "Aeroengines—Altitude Test Beds," p. 221.
22. George W. Lewis, "Report on Trip to Germany and Russia: September-October 1936," *Historical Collection File 11059* (Washington, DC: NASA Headquarters, 1936).
23. Lewis, "Report on Trip to Germany and Russia."
24. Benjamin Pinkel interview, Cleveland, OH, by Virginia Dawson, 4 August 1985, NASA Glenn History Collection, Oral History Collection, Cleveland, OH.
25. Virginia Dawson, *Engines and Innovation: Lewis Laboratory and American Propulsion Technology* (Washington, DC: NASA SP–4306, 1991).
26. Harold Friedman interview, Cleveland, OH, 2 November 2005, NASA Glenn History Collection, Oral History Collection, Cleveland, OH.
27. Robert S. Arrighi, *Revolutionary Atmosphere: The Story of the Altitude Wind Tunnel and the Space Power Chambers* (Washington, DC: NASA SP–2010–4319, 2010).
28. Pinkel interview, 1985.
29. "Major Research Facilities of the Lewis Flight Propulsion Laboratory, NACA, Cleveland, Ohio, Engine Research Building," 17 July 1956, NASA Glenn History Collection, Cleveland, OH.
30. Pinkel interview, 1985.
31. Carlton Kemper, "Dr. Kemper's Talk at Morning Session of First Annual Inspection," *Annual Inspection at NACA Lewis Flight Propulsion Laboratory, 8–10 October 1947*, NASA Glenn History Collection, Cleveland, OH.
32. NACA, "Thirty-Eighth Annual Report of the National Advisory Committee for Aeronautics" (Washington, DC: Government Printing Office, 1954), p. 4.
33. Robert Walker interview, Cleveland, OH, 2 August 2005, NASA Glenn History Collection, Oral History Collection, Cleveland, OH.
34. Howard Wine interview, Cleveland, OH, 4 September 2005, NASA Glenn History Collection, Oral History Collection, Cleveland, OH.
35. Walker interview, 2005.

Image 13: Construction of PSL No. 1 and 2 on 25 January 1951. The two primary coolers for the altitude exhaust are in place within the framework near the center of the photograph. The Shop and Access Building is being built to the left. (NASA C–2009–00180)

· 2 ·
The Next Big Thing

Edward "Ray" Sharp, the Director and congenial father figure of the NACA Lewis Flight Propulsion Laboratory, was a bit tense as he rode downtown on 3 October 1952 to pick up Congressman Albert Thomas at the Hotel Carter. The Texas representative, considered to be "the locus of opposition to the [NACA] in the 1950s," was in town to personally assess the Cleveland laboratory for waste and duplication of services.[1]

Congress passed the Budget and Accounting Procedures Act of 1950 to reduce federal spending in order to fund the Korean War. There was pressure to trim the federal deficit and reduce spending on research and development. Thomas, Chairman of the Appropriation Subcommittee, had halved the NACA's 1950 request for its Unitary Plan wind tunnels, blocked an effort to reestablish a European office, and verbally battled with NACA Chairman Jerome Hunsaker during a 1951 appropriations hearing.[2]

Early in 1952 Thomas initiated a formal assessment of the NACA. A team of young auditors was sent to Lewis in September to scour the financial, property, and personnel records.[3] Now Thomas was in town to make a personal inspection of NACA's Cleveland lab.[4] Sharp noted that, upon exiting the vehicle, Thomas immediately commented on the cracks in the Administration Building facade.[5] It figured to be a long day.

Once inside, Sharp called upon his brilliant technical partner, Abe Silverstein, to assist. Sharp managed the lab operations, but Silverstein directed the research activities. The two confidants fielded the Congressman's polite, but multitudinous, inquiries as they explored the lab throughout the morning.[6]

Image 14: Abe Silverstein (left) and Ray Sharp (right) host a more relaxed tour of the laboratory in 1953. (NASA C–1953–34030)

Pursuit of Power

At the end of the tour, Sharp and Silverstein made a point of taking Thomas to their newest facility, the Propulsion Systems Laboratory (PSL), where technicians were preparing for the facility's very first tests. There Silverstein explained the necessity for testing full-size engines in simulated altitude conditions, and the benefits of remedying problems before the engines went into production. Sharp later remarked, "I think [Thomas] realized for the first time the potentialities of this work since he commented that it was obvious that large savings could be made if the engines were proven before they were put into production."[7]

The cantankerous Congressman left Cleveland with a favorable impression. The PSL provided a tangible real-world example of how NACA research could benefit the nation. The lab had a number of unique facilities, but the PSL was the latest and most influential. It was an investment that would pay dividends for years to come—an investment that had begun five years earlier in 1947.

NACA's Panel of Experts

The performance of jet engines began to increase dramatically in the years following the end of World War II. It was clear that the Cleveland lab would have to invest in new engine test facilities to stay ahead of the curve. A new "high-use-factor" facility was needed for fundamental research on engines for aircraft and missiles.[8]

A Research Facilities Panel was assembled to determine the requirements and infrastructure needed for this new engine facility. The panel included several of the lab's most influential leaders: Silverstein, the wind tunnel expert; Ben Pinkel, the thermodynamicist; Oscar Schey, the compressor and turbine authority; John Collins, engine performance specialist; and Carlton Kemper, the executive engineer.[9]

NACA engineers knew that they wanted to study larger engines in a chamber that could recreate flight

Image 15: Back row, left to right: Abe Silverstein and Oscar Schey. Middle row, left to right: Benjamin Pinkel, Jesse Hall, and John Collins. Ray Sharp is in front of Hall and Collins. Pinkel, Silverstein, Schey, Collins, and Carlton Kemper (not pictured) were members of the Research Facilities Panel. (NASA C–1949–22821)

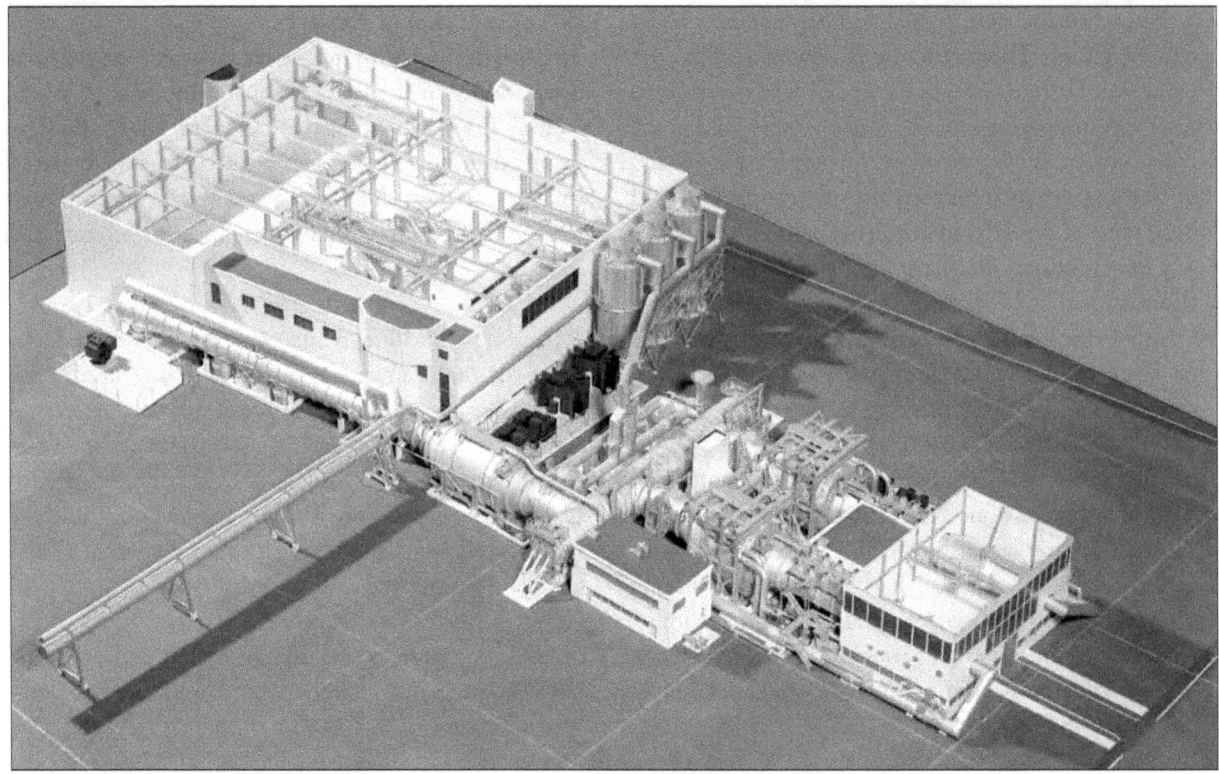

Image 16: Model of PSL No. 1 and 2. The Equipment Building is to the left, and the Shop and Access Building housing the test cells and control room is to the right. (NASA C–1954–37139)

conditions, but many questions remained unanswered: How powerful would future engines be? Exactly what conditions did the facility have to simulate? What infrastructure and systems would be needed to accomplish the simulation?

The panel convened on 3 November 1947 to discuss these issues. Pilot Chuck Yeager had recently broken the sound barrier at Muroc Lake, and the Four Burner Area was about to begin operation at the Cleveland lab. The panel, under Pinkel's leadership, deemed three elements to be necessary: the ability to expand in the future, the proximity to electrical supplies, and the ability to operate around the clock seven days a week. The group was able to establish the broad parameters that would guide the design team.[10]

Congressional Authorization

The NACA had requested more than $12 million for the PSL construction.[11] In 1948 Congress authorized $10 million for the project, including $3 million for architectural and engineering services.[12] It appears that the timing of the budget request was perfect. If the management at NACA's Cleveland lab had delayed even a year or two, the request might have been rejected.

As previously mentioned, the lab was entering a difficult phase brought on by the onset of the Korean War in June 1950. World War I led to the creation of the NACA, and World War II spurred the establishment of two new NACA laboratories. The Korean conflict, however, resulted in cutbacks across the federal government, including a three-year budgetary regression for the NACA. Ninety NACA Lewis Flight Propulsion Laboratory[c] employees were laid off in January 1952, and budgets for new facilities and research programs were reduced.[13] The PSL was largely spared from the cutbacks because its funds had been approved before the war.

On 20 November 1950 the commercial architects estimated PSL's total cost of construction at roughly $11.8 million. By far the largest line item was the $1.872 million Exhauster System.[14]

A two-phase approach was implemented to expedite PSL's construction. The first phase sought to have the facility operating by 1 January 1951 in order to test existing engines. The second phase would upgrade its capabilities to accommodate the larger engines of the future. All elements of the design had to be adaptable to this future expansion.[15]

The $10 million Phase I included the combustion air supply, altitude exhaust, and process water systems.[16] The PSL was designed to simulate altitudes of 50,000 feet and to handle engines of up to 15,000 pounds of sea-level thrust, which was more thrust than produced by any engines available in 1948.[17] Phase II would increase the combustion air capability to simulate an altitude of 80,000 feet and add a refrigeration system. Funds for this phase were approved in July 1952 at the height of Congressman Thomas's crusade against the NACA.[18] PSL's capabilities were further improved over the years.

The Designers

The PSL was composed of three major components: a combustion air system, an altitude exhaust system, and altitude tank equipment. The basic concept was relatively simple. Pinkel explained to historian Virginia Dawson, "Well, you start off with a chamber. You logically take a cylindrical shape. You then need enough room for the man to work in it. You need a cover that opens up. You need to set it up on some sort of stand where you can measure thrust of the engine.... Then you need the compressors and coolers that you need to simulate the altitude conditions. In spite of the fact that it looks impressive, it is a pretty straightforward thing."[19]

Eugene Wasielewski was responsible for converting Pinkel's vision for the PSL into engineering specifications. With the assistance of Dan Williams, Bruce Lundin, and refrigeration specialist Achille Gelalles, Wasielewski was able to accomplish the bulk of PSL's engineering work within five months. Construction Administrator James Braig's staff then provided the detailed design work. Once Wasielewski approved, the construction specifications were put out for bids.[20] Local architectural firm Burns and Roe, Inc., was hired to expedite the final design process.[21]

Despite the overall simplicity of the altitude tank concept, the specifics of the PSL design were complex. Burns and Roe worked with NACA Lewis engineers to continually modify and update the plans as the project progressed.[22] Other specialists were brought in to create various components. The compressors were designed by the Elliott Company, the exhausters by Roots-Connersville Corporation, the primary and secondary coolers by Ross Heater Company, and the two altitude chambers by Treadwell Construction.[23,24]

The struggle to develop a thrust-measuring mechanism for the test sections was one example of the difficulty of the final design. The original thrust stand design was unsatisfactory, but the engineers were unable to develop an alternative. After much debate, the existing design was used with several modifications.[25]

Burns and Roe promised to complete the design work by May 1950, but difficulties in coordinating the bevy of contractors delayed the effort.[26] Nonetheless Lewis Director Ray Sharp was pleased with the company's work when it concluded in January 1951.[27]

Eugene Wasielewski

Wasielewski arrived at the NACA's Cleveland lab in 1947 and quickly established himself as one of the NACA's central figures of the 1950s. Almost immediately, he was tasked with the design of the most powerful altitude tanks in the nation, the PSL. Afterward he became a leader of the design team for the 10- by 10-Foot Supersonic Wind Tunnel, the most powerful propulsion wind tunnel in the nation.

Wasielewski earned his mechanical and aeronautical engineering degree (1934) and a masters in engineering mechanics (1935) from the University of Michigan. Upon graduation, he took a position at engine manufacturer Allis-Chalmers. By 1937, however, Wasielewski had joined the NACA at Langley Field.[28]

There Wasielewski and Eastman Jacobs undertook an effort to improve General Electric's supercharger. They employed a new type of compressor design—an eight-stage axial-flow compressor. Although the axial-flow design did not prove practical for the supercharger application, it would influence the future design of jet engines in the United States.[29]

Wasielewski left Langley in 1941 for a position at Ranger Aircraft Engines. He rejoined the NACA in 1947 at its Cleveland lab.[30] There he was a crucial figure in the design of the lab's two major projects of the 1950s—the PSL and the 10- by 10-Foot Supersonic Wind Tunnel.

In just a few years Wasielewski was promoted to the highest ranks at the NACA Lewis Flight Propulsion Laboratory. He was named Chief of the Engine Research Division in 1949, Assistant Chief of the Technical Services Division in November 1952, Chief of the Unitary Plan Activity in 1954, and Assistant Director of the entire lab in 1955. He retired the following year.

Wasielewski had left the NACA to design the Altitude Facility at Curtiss-Wright Corporation's Wood-Ridge, New Jersey, site. Three years later he was named Curtiss-Wright's Chief Engineer. In that role he helped design additional test facilities and engines.[31]

In October 1960 Wasielewski returned to government service at NASA's newly established Goddard Space Flight Center. Abe Silverstein, Wasielewski's former boss at Lewis, was then serving at NASA Headquarters. Silverstein not only directed all spaceflight programs but planned the Goddard laboratory and served as its first Director.

Wasielewski came on board shortly thereafter and served, until his second retirement in 1967, as the principal institutional manager for the new Director, Harry Goett.[32] Wasielewski, his wife Virginia, and their five children permanently relocated to the Washington, DC, area. He passed away in 1972 at the age of 60.

Image 17: Wasielewski at the opening of the 10- by 10-Foot Supersonic Wind Tunnel on 22 May 1956. (NASA C–1956–42127)

Propulsion Systems Laboratory Rising

Image 18: These two stands were erected to hold PSL's two primary coolers on 26 April 1950. Construction at the site had been under way for approximately six months at this point. (NASA C–2009–00175)

Image 19: H.K. Ferguson Company constructs an overhead air pipe for the PSL site in August 1949. The pipe joined PSL's combustion air system with the Altitude Wind Tunnel (AWT) and Engine Research Building. (NASA C–2009–00181)

Image 20: Future site of the PSL as it appeared in October 1949. A process air line is seen being built from the Engine Research Building to the PSL site. The AWT is near the upper right corner of this photograph. (NASA C–1949–24972)

Image 21: Early phases of construction in late December 1949. The steel structure to the left would support one of the large exhaust coolers. The shell of one of the coolers is sitting in the background. By the early spring of 1950, the Shop and Access Building was up, and the massive steel stands for the two primary coolers were in place behind. (NASA C–2009–00151)

Image 22: Excavations for the Equipment Building were begun in June 1950. This photograph shows the progress in February 1951 (NASA C–2009–00183)

Pursuit of Power

Image 23: Installation of one of the primary coolers in June 1950. When operational, the vertical fins were filled with cooling water that removed heat from the exhaust flowing through the fins. The cooling water carried the heat to the cooling tower where it was dissipated. (NASA C–2009–00177)

Image 24: Early stages of construction of the Shop and Access Building in August 1950. The two altitude chambers and control room would be added to the second floor later. (NASA C–2009–00179)

Image 25: This late September 1950 photograph shows the framework for the two primary coolers for the altitude exhaust near the center of the photograph. The Shop and Access Building is being built to the right, and excavation for the Equipment Building has begun in the background to the left. (NASA C–2009–00184)

Image 26: A caravan of large steel castings manufactured by the Henry Pratt Company arrived on 5 January 1951 for the PSL altitude chambers. The massive pieces were delivered to the area by rail and then loaded on a series of flatbed trucks that transported them to Lewis. The nearest vehicle has one of the clamshell access hatches. (C–2009–00173)

Image 27: The inlet section of one of the PSL altitude chambers is unloaded from a flatbed at the construction site on 5 January 1951. Air would enter the tank through an opening on the right side during operation. (NASA C–2009–00171)

Image 28: The Shop and Access Building during construction in May 1951. The two portals on the front would later permit conditioned air to be piped into the two altitude chambers inside. (NASA C–2009–00195)

Image 29: Construction of the PSL Equipment Building viewed from the north in July 1951. This structure housed the exhausters and compressors. (NASA C–2009–00166)

Image 30: Elliott Company compressors being installed in the Equipment Building during April 1952. The Roots-Connersville exhausters can be seen on the far side of the room near the top of the photograph. (NASA C–1952–29500)

Image 31: Construction of the gas-fired air heaters outside the Equipment Building on 29 August 1951. (NASA C–2009–00163)

Image 32: Construction of the wooden PSL Cooling Tower in a wooded area across the road from the PSL in September 1949. (NASA C–2009–00161)

Executing the Plans

Wasielewski and James Braig segregated the construction into four areas: electrical and power systems, mechanical equipment, research and control equipment, and utilities and infrastructure.[33]

Construction began late in the summer of 1949, although the design work would stretch out almost a year and a half longer. By the summer of 1951 the structure of the Shop and Access Building was nearly complete. The intercooler and air heaters were installed in August, and work on the Equipment Building and Circulating Water Pump House progressed over the next year. The compressors and exhausters were installed in the spring of 1952.[34–37]

The exhaust, compressor, and other systems were calibrated throughout the spring and summer of 1952. The airflow and altitude limits had to be determined before any actual tests were run. Bob Walker explained, "On one end you had the big after-cooler and exhaust system…on the other end you had the compressors that were blowing the air in the front. You tried to establish the airflow and altitude that would be necessary for whatever test you wanted to run. We calibrated the whole facility like that."[38]

Assembling the Crew

As the construction of the PSL progressed, NACA Lewis began making arrangements to staff the facility and integrate it into the lab's existing framework. In July 1951 Lewis's Engine Research Division was restructured to incorporate the new PSL facility. The division included Lewis's top veteran propulsion experts. Gene Wasielewski was named division chief and Bruce Lundin his assistant. Researchers were scattered over several branches. AWT Chief Alfred Young was selected to oversee the division's technical operations at both the AWT and PSL, with assistant Elmo Farmer directly responsible for the PSL. The division was responsible for initiating the tests and documenting their findings in technical reports.[39]

Bill Egan's Test Installations Division was responsible for preparing the test equipment, installing it in the facility, and sometimes running the actual tests. The no-nonsense section head Erwin "Bud" Meilander oversaw all of the PSL mechanics and was responsible for making the researchers' visions into a reality in the PSL test chambers. Paul Rennick and Carl Betz were the day-shift supervisors, and Lyle Dickerhoff was "the unseen member" running the tests at night.

Images 33–35: Left to right: Bud Meilander, Lyle Dickerhoff, and Carl Betz. (NASA C–1966–01760, NASA C–1958–48725, and NASA C–1956–42477)

In October 1952, almost five years after it was first envisioned, the PSL was ready to tackle the propulsion problems of the Cold War. It was also ready for Congressman Thomas's 3 October visit. The Cold War was at the forefront of Thomas's mind at the time. During his tour he not only talked about cost savings but about the nation's standing with the Soviet Union. He was particularly interested in the field of long-range missiles.[40] At the PSL, technicians were finalizing the setup to study the powerful XRJ47 ramjet for the Navaho Missile program.

Image 36: PSL No. 1's primary cooler as it appeared on 24 September 1952, just after completion. After months of calibration, the PSL was about to commence testing. (NASA C–1952–30767)

Endnotes for Chapter 2

1. Alex Roland, *Model Research*, Volumes 1 and 2 (Washington, DC: NASA SP–1403, 1985), chap. 11.
2. Roland, *Model Research*.
3. Eugene Braig, "Memorandum for Files, General Accounting Office Survey and Investigation of the NACA Lewis Laboratory," 5 November 1952, Glenn History Collection, Dawson Collection, Cleveland, OH.
4. Roland, *Model Research*.
5. Edward Sharp, "Memorandum for Director, NACA, Visit of Congressman Albert Thomas and Mr. A. H. Skarin, Friday, Oct 3, 1952," Glenn History Collection, Directors' Collection, Cleveland, OH.
6. Sharp, "Memorandum for Director, …Visit of Congressman Albert Thomas…."
7. Sharp, "Memorandum for Director, …Visit of Congressman Albert Thomas…."
8. Carlton Kemper, et al., "Memorandum for Director, Report of the Special Panel Appointed to Study General Requirements of the Propulsion Sciences Laboratory," 18 March 1948, NASA Glenn History Collection, Test Facilities Collection, Cleveland, OH.
9. "Cleveland—Lewis" organizational list, 30 September 1948, NASA Glenn History Collection, Cleveland, OH.
10. "Cleveland—Lewis" organizational list, 1948.
11. Jesse Hall, "Division Chief Staff Conference," 3 November 1947, NASA Glenn History Collection, Cleveland, OH.
12. "Design Specifications for Architect-Engineer Services for the Propulsion Sciences Laboratory Equipment Building, Project No. 794," 28 November 1949, NASA Glenn History Collection, Test Facilities Collection, Cleveland, OH.
13. Roland, *Model Research*.
14. Burns and Roe, Inc., "Estimated Cost of Propulsion Science Laboratory as of November 20, 1950," NASA Glenn History Collection, Test Facilities Collection, Cleveland, OH.
15. "Specifications for Furnishing Architect-Engineer Services for the Propulsion Sciences Laboratory Phase I Project No. 794, NACA Lewis Propulsion Research Laboratory," 5 August 1948, NASA Glenn History Collection, Test Facilities Collection, Cleveland, OH.
16. "Specifications for Furnishing Architect-Engineer Services."
17. Kemper, et al., "Memorandum for Director, …Special Panel…Propulsion Sciences Laboratory."
18. C. S. Moore, "Appendix A, Proposal for Propulsion Sciences Laboratory," NASA Glenn History Collection, Directors' Collection, Cleveland, OH.
19. Benjamin Pinkel interview, Cleveland, by Virginia Dawson, 4 August 1985, NASA Glenn History Collection, Oral History Collection, Cleveland, OH.
20. Edward R. Sharp, "Appointment of Propulsion Sciences Laboratory Project Engineer," 17 March 1948, NASA Glenn History Collection, Directors' Collection, Cleveland, OH.
21. Kemper, et al., "Memorandum for Director, …Special Panel…Propulsion Sciences Laboratory."
22. Edward R. Sharp to R.C. Roe, "Contract NAw-5652, Architect Engineer Services for Propulsion Systems Laboratory," 29 December 1950, NASA Glenn History Collection, Directors' Collection, Cleveland, OH.
23. Burns and Roe, Inc., "Progress Report No. 22 for Propulsion Science Lab Phase I Part II," 6 October 1949, NASA Glenn History Collection, Test Facilities Collection, Cleveland, OH.
24. Burns and Roe, Inc., "Progress Report No. 20 for Propulsion Science Lab Phase I Part II," 8 September 1949, NASA Glenn History Collection, Test Facilities Collection, Cleveland, OH.
25. Burns and Roe, Inc. "Progress Report No. 24 for Propulsion Science Laboratory Phase I Part II," 2 November 1949, NASA Glenn History Collection, Test Facilities Collection, Cleveland, OH.
26. Burns and Roe to Contracting Officer at NACA Lewis Flight Propulsion Laboratory, "Contract NAw-5652, Architect Engineer Services for Propulsion Systems Laboratory," 15 March 1950, NASA Glenn History Collection, Test Facilities Collection, Cleveland, OH.
27. Sharp to Roe, "Contract NAw-5652."
28. "Eugene W. Wasielewski Named Associate Director of Goddard Space Flight Center," *Goddard News*, 17 October 1960.
29. James Hansen, Engineer in Charge: A History of the Langley Aeronautical Laboratory, 1917–1958 (Washington, DC: NASA SP–4305, 1987), chap. 8.
30. "Wasielewski Honored at Farewell Party," *Wing Tips* (12 September 1966).
31. "Wasielewski Named Associate Director of Goddard Space Flight Center."
32. "Wasielewski Named Associate Director of Goddard Space Flight Center."

33. Edward R. Sharp to Staff, "Organization and Responsibilities of the Staff of Project Engineer, Propulsion Sciences Laboratory," 14 June 1948, NASA Glenn History Collection, Test Facilities Collection, Cleveland, OH.
34. Burns and Roe, "Progress Report No. 20 for Propulsion Science Lab Phase I Part II."
35. "Construction of Propulsion Systems Laboratory (PSL) Shop and Access Building," May 1951, C–2009–00195. NASA Glenn Imaging Technology Center, Cleveland, OH. (Image 28).
36. "Construction of Cooling Tower for Propulsion Systems Laboratory (PSL)," 1950, C–2009–00186, NASA Glenn Imaging Technology Center, Cleveland, OH.
37. "Elliott Company Compressors Being Installed in the Equipment Building," Apr. 1952, C–1952–29500, NASA Glenn Imaging Technology Center, Cleveland, OH. (Image 30).
38. Robert Walker interview, Cleveland, OH, 2 August 2005, NASA Glenn History Collection, Oral History Collection, Cleveland, OH.
39. Abe Silverstein, memorandum, "Organizational Changes Within the Engine Research Division," 26 July 1951, Glenn History Collection, Directors' Collection, Cleveland, OH.
40. Sharp, "Memorandum for Director, …Visit of Congressman Albert Thomas…."

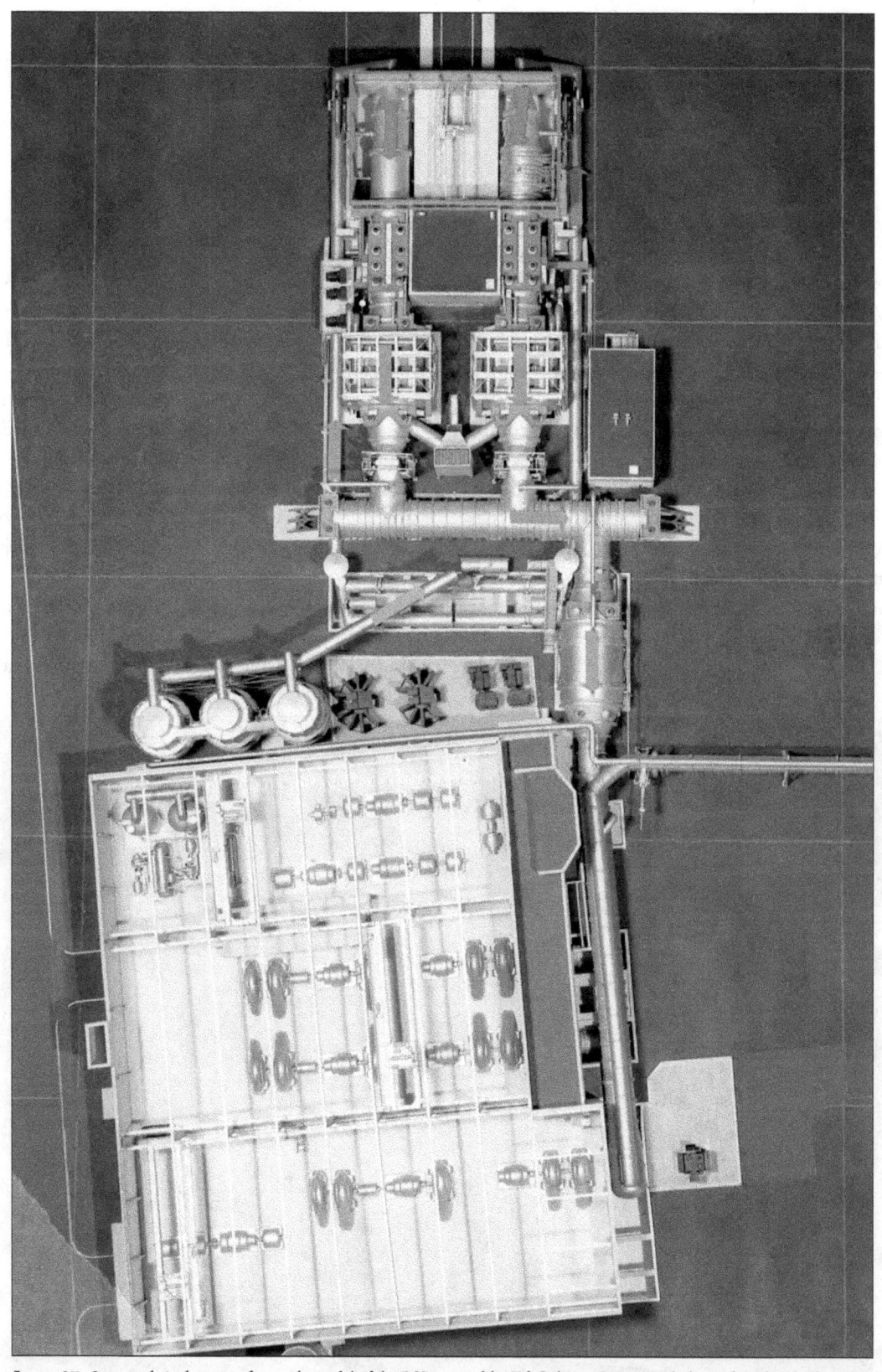

Image 37: One-tenth-inch to one-foot scale model of the PSL created by Ed Calmer, George Perhala, and Dick Schulke. The model, seen here on 19 November 1954, amalgamated all the drawings from various contractors for the first time.[1] This model was used in the NACA Lewis 1954 Inspection. A film was created afterward using arrows to demonstrate the airflow pattern. (NASA C–1954–37141)

· 3 ·
Harnessing PSL's Muscle

Three spotlights glared down on Martin Saari as he stood before 60 of the nation's most prominent aerospace professionals and a forest of missile and turbojet models. The group listened intently as the propulsion research veteran explained the importance of full-scale engine testing—engine components may seem to operate flawlessly on their own, but they can manifest problems when integrated with other engine components. Saari then described the direct-connect and free-jet methods used to study full-scale engine systems: "The newest and largest facility in which these two techniques are used is the Propulsion Systems Laboratory [PSL] in which you are now seated," he explained as he strode across the stage to a mounted model of the facility.[2]

The crowd, seated between the two open test chambers, took note as Saari indicated the chambers on the cutaway model. The PSL was the state of the art—the most advanced facility in the nation for studying full-scale engine systems, and the NACA's Triennial Inspection in June 1954 was the first time that the larger aerospace community was seeing it. Every year, one of the three NACA labs opened its doors to industry representatives and the military for an "Inspection." It was the NACA Lewis Flight Propulsion Laboratory's turn in 1954, and the visitors were eager to hear how the unique new facility worked.

PSL's immense air-handling system differentiated the PSL from other altitude facilities like the Altitude Wind Tunnel. Moving his pointer to a second, larger

Image 38: Seymour Himmel at the PSL display for the NACA Inspection at Lewis on 17 June 1954. The previous speakers, Martin Saari and Bill Conrad, had explained how the PSL worked using the facility model hanging in the upper left corner of this photograph. (NASA C–1954–36018)

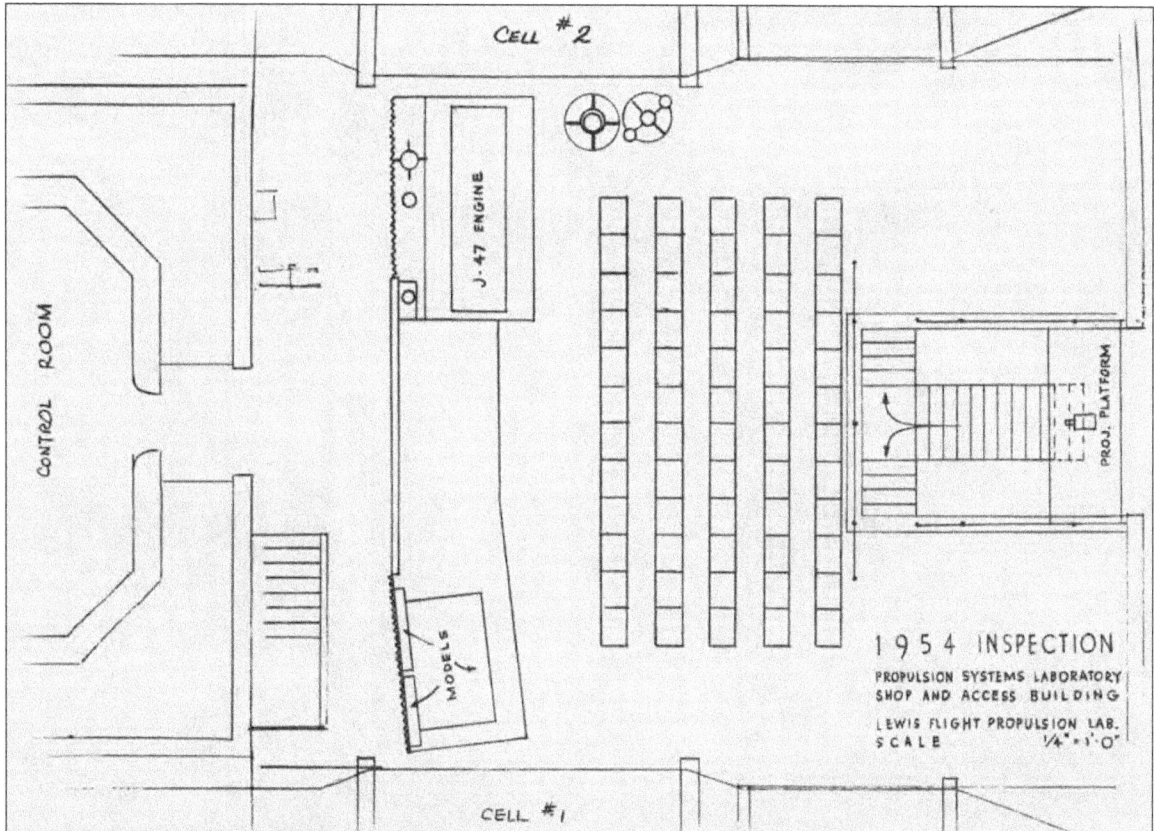

Image 39: Layout of the 1954 Inspection presentation held between the PSL's two test chambers. The display ran between the two test sections with rows of chairs in front. The control room to the left was behind the display. (NASA)³

building on the model, Saari explained, "All of this remaining equipment is used to provide the proper environment for the test engines. The air is delivered at high pressure by several compressors located in this Equipment Building. Also located in this building are air driers, refrigeration turbine and air heaters. The low pressure in the test chamber required for altitude simulation is provided by several large suction pumps or exhausters in this building."[4]

The visitors could not view the control room, which was behind the stage, but the chambers had their hatches open so that the engines, thrust stands, and extensive instrumentation were on display.

During the tests, the engines generated large quantities of high-temperature gases that had to be mitigated. Saari pointed to the two massive cylinders behind the test chambers and explained, "The extremely hot exhaust gases from the engine are cooled by these large coolers before passing into the exhausters."[5]

In just a few sentences, Saari had explained how the nation's most modern engine research facility was run. Of course, there was more to conducting a test than just powering up the facility.

Making Research a Reality

Conducting a test program required a great deal of coordination between the researchers, the mechanics, the technicians, and the technical engineers. There were a few rivalries between some of the groups, but they were friendly on a personal basis and dedicated to making the test program succeed.

The research divisions generally consisted of different specialized branches of researchers with a separate operations branch for test and electrical engineers. The research engineers would work with the test engineers to determine the overall objectives for the test. The researchers would tell the test engineers what controls, set points, and instrumentation they wanted. The test engineers would design the test plan, work with

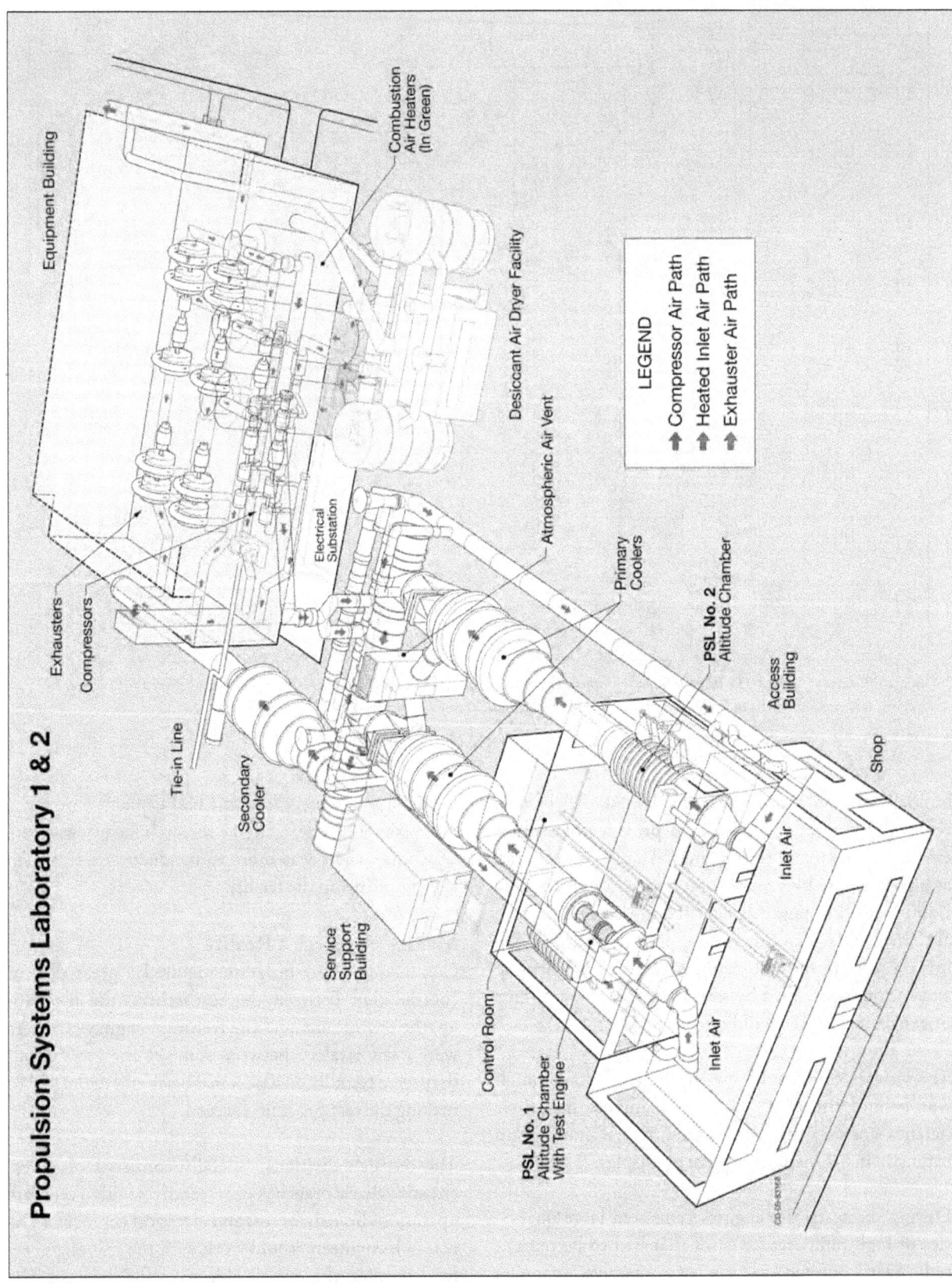

Image 40: Layout of PSL No. 1 and 2. The Shop and Access Building is at the front left, and the Equipment Building is behind to the right. Arrows indicate the airflow. (2009; NASA CD–09–83168)

the technicians and electricians to install it, and verify that it worked properly.[6]

The mechanics, electricians, and electronics technicians from the Test Installation Division were responsible for maintaining the facility and installing the equipment for the tests. The mechanics had their own supervisors within the Test Installations Division, but they worked closely with the research divisions.

"It's one of the strange things about the setup of the lab here," Neal Wingenfeld recalled. "They work with us [research engineers] all day long, these technicians. We told them what to do. When they had technical difficulties they came to us. Their supervisor came around and checked on them. Told them how much sick leave and annual leave they had. They filled out their time sheets, and gave them their safety talks, but other than that, they worked for us. So we called ourselves their technical supervisors, and actually 99 percent of the time we told them what to do. We set up their schedules, we gave them their workloads, we supplied their Kleenex, their tools, and their technical help, and trained them."[7]

The test engineers were also responsible for managing and running the actual tests. For each day of testing, the test engineer had to coordinate with the mechanics and technicians, the lab's data acquisition team, the operators in the Equipment Building, the electric company engineers, those responsible for the fuel supply, and any representatives from the engine manufacturer.[8]

The tests were run overnight, so many times the researchers were not present. Thus the mechanics and test engineers often had a much better understanding of the tests than did the researchers, who were debriefed first thing in the morning.

"Well, some of the documents could be somewhat misleading," explained Howard Wine. "You had many, many instances where the engineers [would] come in with a proposal to run some tests that night and they didn't get the results they wanted for one reason or another, and they say, 'well, we've got to change this now, through this nozzle-area or do something.'... And we [mechanics] would always say, 'well, why don't you go back to the office and come back at five and maybe we'll come up with something.' So they would go back to their office, and we would start getting the hammers and cutting torches and welding machines and everything cranked up and make the modifications. And they would come back and say, 'that's great, now wait a minute, let me copy that down so I [can] put that on the drawing.' You know so sometimes the drawing came after the fact."[9]

NASA's Propulsion Systems Laboratory No. 1 and 2

· Conducting a Test at the Propulsion Systems Laboratory ·

Image 41: Research engineers, such as Martin Saari (on the left), responded to requests submitted to the NACA by the military or engine manufacturers by developing a suite of tests for the particular engine. The engineers then determined the proper test facility to conduct the investigations. In the 1950s, Lewis had three facilities for conducting full-scale engine tests, but by the 1960s, the PSL had become the only facility in the Agency capable of performing this type of work. (NASA C–1957–45797)

Image 42: The Test Installations Division would review the researcher's request, coordinate with the manufacturer to obtain the engine and equipment, and fit the test into the facility's schedule. The engine would then be shipped to the facility and instrumented for the test in the first-floor shop area, seen here. Thermocouples, pressure rakes, and other measurement equipment were prepared in the first-floor Instrument Shop. (NASA C–1967–01582)

Harnessing PSL's Muscle

Image 43: An overhead crane transported the engine to the second floor where final preparations were made before it was placed into one of the two chambers through the chamber's large clamshell hatch. (NASA C–1955–40070)

Image 44: Test engineers and mechanics installed the engine and altered the configuration for the particular type of test. This required coordination between the researcher, facilities engineers, and technicians. (NASA C–1957–45798)

Images 45–46: The PSL could be configured in either a direct-connect or free-jet mode. Left: The direct-connect was the simplest way of studying the internal performance of the engine. The engine was mounted on the thrust stand with the airflow connected directly to the engine inlet. (NASA C–1955–40235) Right: The free-jet setup was more beneficial because the entire engine system, including the inlet duct, could be studied. A nozzle was used to create a supersonic jet of air that enveloped the engine inlet in supersonic altitude air. (NASA C–1954–36894)

Image 47: To obtain useful data from the tests, technicians and electricians had to place instrumentation in both the engine and the test chamber. It could take weeks or even months to install the multitude of thermocouples, rakes, and other instruments. Sometimes each compressor stage had to have its own readings. The engine itself was atop a thrust stand that measured the thrust and drag. In this photograph, mechanics check the instrumentation on a J85 turbojet on 19 August 1977. (NASA C–1977–03128)

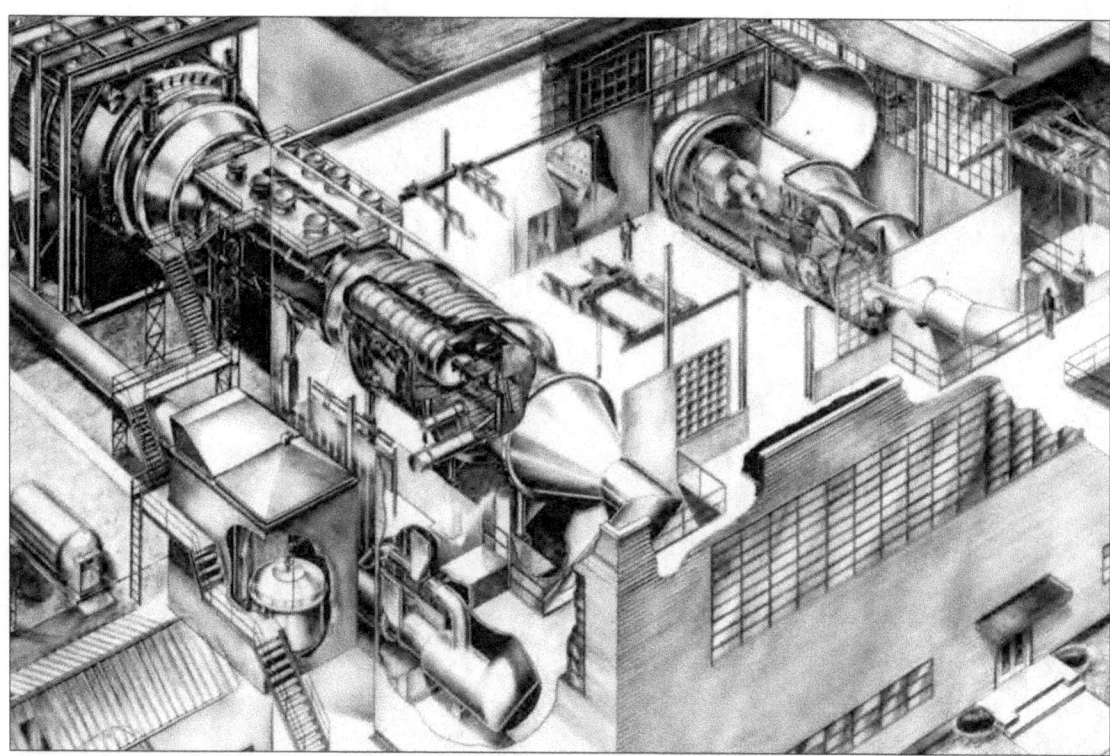

Image 48: Cutaway of the second floor of the Shop and Access Building where the technical staff spent most of its time. The test chambers, control room, and tool crib were located in the area. (NASA 8090EL)

Image 49: The first and second shift worked with the research engineers to repair any damage from the previous night's test, make modifications to the setup, and prepare for the next test runs. The mechanics frequently rectified problems with the test program. (NASA C–1953–33436)

Image 50: The third shift would come in overnight to actually run the tests. The facility required so much electricity that the NACA had an agreement with the local electric company to operate the facility only at night. Many of the researchers did not care for the late-night testing and sought out operational engineers to run the tests. (NASA C–1952–30766)

Image 51: "PSL 1 and 2 had lots and lots of pipes and big valves," remembered Howard Wine. "It could take you about an hour just to check the facility out every night prior to running to make sure the valves were all [in the] right sequence....You checked everything, climbed the ladders up to the top, and [made] sure the flow-out disks were all in place and so forth. So it was quite an operation we went through running the facility here."[10] (NASA C–1952–30768)

Image 52: Once the large clamshell lid was closed and the chamber was sealed, the altitude conditions were introduced, and the engine was ignited. This 2008 photograph shows PSL No. 1 with the lid closed. The large gear wheels behind the closed lid raised the large lid mechanically. (NASA C–2008–04129)

Image 53: Facility engineers in the Equipment Building coordinated with the electric company and the main air-handling control room in the Engine Research Building (ERB) basement. The equipment was linked to Lewis's central air system, which allowed the PSL system to augment the compressors in other test facilities. Slowly the facility engineers began bringing all the air-handling systems online. (NASA C–1953–31863)

Image 54: These 16,500-horsepower Elliott Company compressors in the Equipment Building, seen on 7 May 1954, supplied high-speed air to the PSL altitude chamber to simulate flight speeds. (NASA C–1954–35700)

Image 55: After passing through temperature-adjusting equipment, the combustion airflow was pumped through a pipe to one of the test chambers on the second floor of the Shop and Access Building, seen here on 3 February 1953. In 1957 a two-story shop section was added to the front of the building. The pipes remained but were not visible on the exterior. (NASA C–1953–31941)

Image 56: Depending on the type of test being run, the airflow into the engine would have to be either heated or refrigerated. Ramjets operated at supersonic speeds and needed to be tested in hot conditions to simulate the heat generated by their velocity. Ambient air was taken into these two heaters and passed through natural-gas-heated vertical tubes. Each heater warmed 125 pounds of air per second degree Fahrenheit.[11] (NASA C–1952–30936)

Image 57: Jet engines traveling at subsonic speeds had to be tested in cold conditions to simulate the high-altitude air temperature. The air flowed upward through a vertical tank and over cascade trays cooled with cooling tower water and then with Freon-chilled water supplied by the compressors shown in this 21 July 1953 photograph. The air at high altitudes is also very dry, so after the air was cooled, as it made its way to the test section, the airflow passed through an air dryer with 190,000 pounds of activated alumina that absorbed moisture.[12] (NASA C–1953–3228)

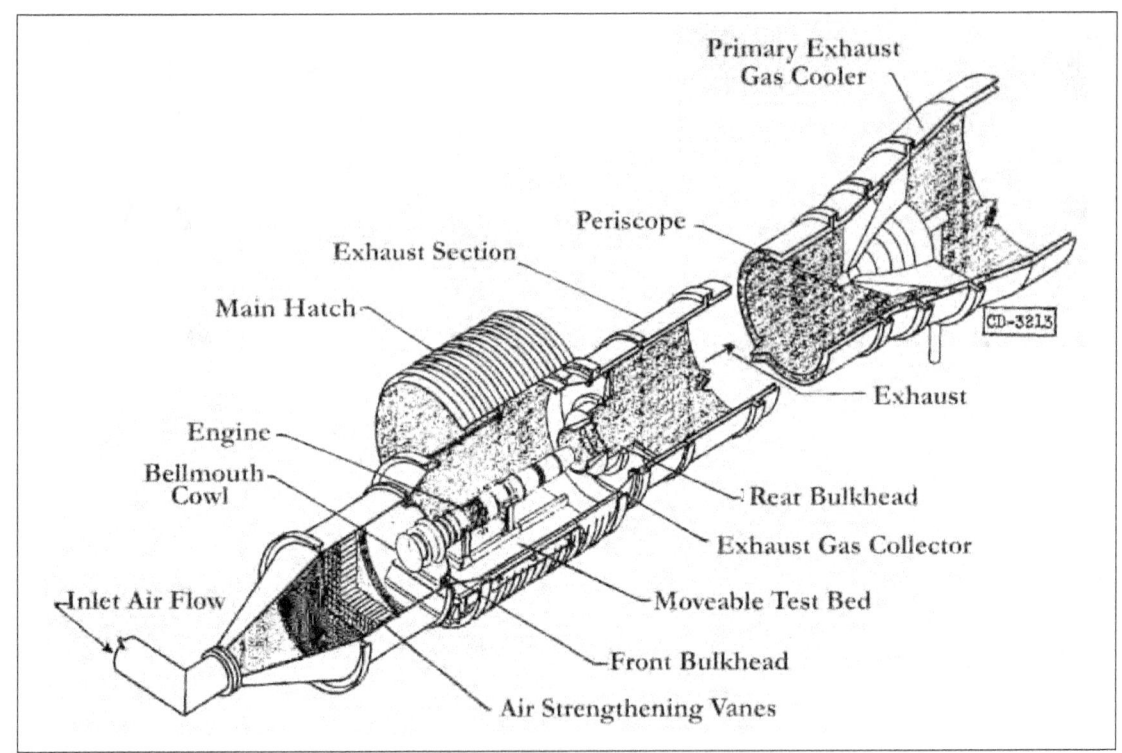

Image 58: The PSL's two 100-foot-long altitude chambers ran parallel to one another inside the Shop and Access Building. Each chamber comprised three sections: the inlet section, test section, and exhaust section. An engine platform inside the test section held the engine and measured its thrust loads and drag. (NACA RM E53I08, figure 3, 1954)

Image 59: Overall view of PSL No. 2 on 16 October 1952. Vanes in the section to the right straightened the airflow as it entered. The flow was accelerated to supersonic speeds through a bellmouth cowl in the bulkhead then passed through the engine in the test section. The engine's exhaust was ejected to the left into a cooler outside of the building. (NASA C-1952-30962)

Image 60: Howard Wine (on the left) monitors temperature readings on 26 July 1957. From these control panels, operators could run an engine in PSL No. 2 at various speeds and adjust the altitude conditions to the desired levels. The center station controlled the engine, and the controls at the right monitored the air-handling and altitude systems. (NASA C–1957–45647)

Image 61: PSL Crew Chief Bob Walker at the control board (foreground) on 9 June 1954. Banks of manometer boards are set up in the rear of the room. Cameras on pedestals photographed the manometers to record the different dynamic pressure readings for later analysis. Obtaining the desired test conditions could be a difficult process; once they were achieved, the operators would continue running the test as long as possible.[13] (NASA C–1954–35900)

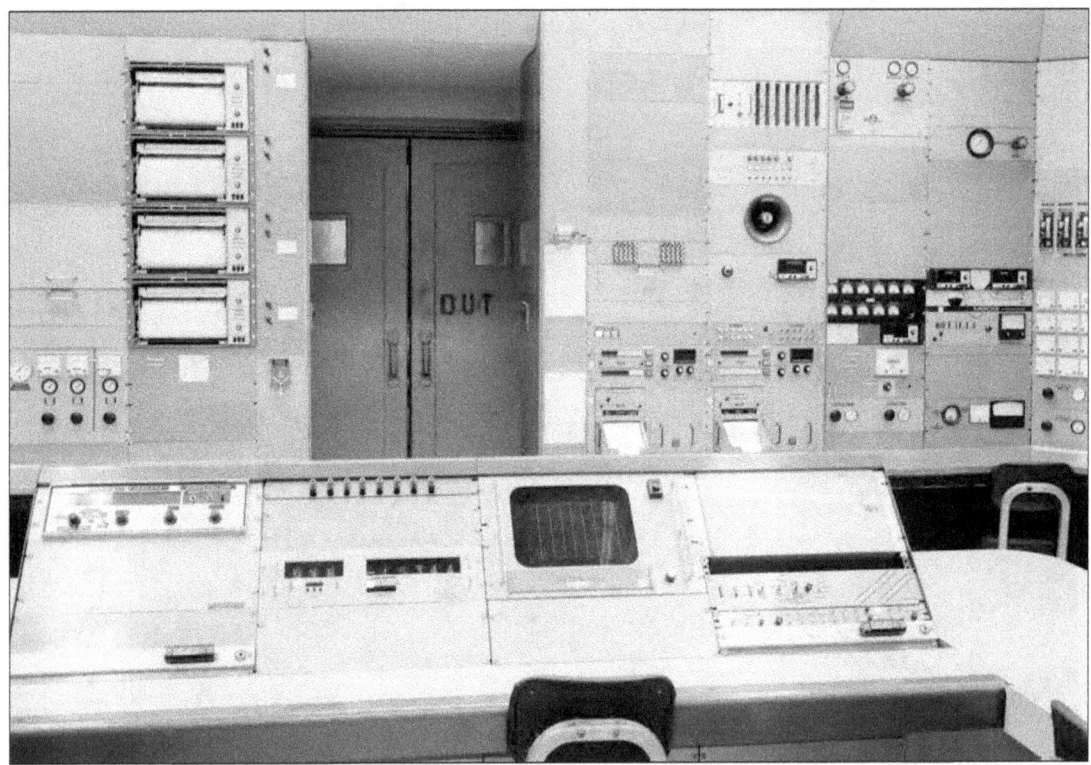

Image 62: The control room was located between the PSL No. 1 and 2 chambers on the second floor of the Shop and Access Building. The chambers were located just through the doors shown in this 12 April 1973 photograph. The control room remained relatively quiet during the tests despite the large engines running just outside its doors. (NASA C–1973–01519)

Image 63: Entrance to the control room. In the foreground, an operations engineer inspects the setup for an engine test in PSL No. 2. Former electrical engineer Neal Wingenfeld recalled how surprisingly quiet it was, "Even though we were sitting only fifty feet from the actual engine you didn't hear it…not the engine running."[14] (NASA C–1963–63291)

Image 64: The large Roots-Connersville exhauster equipment thinned the air in the test chambers to simulate high altitudes, and it removed hot gases generated by the test engines. "All that equipment you see in the PSL Equipment Building, every one of those exhausters was running. If you really needed it they had a tie line to the ERB, and they would start everything in the ERB basement," remembered Walker.[15] The No. 43 exhauster is being repaired in this 6 January 1959 photograph. (NASA C–1959–49428)

Image 65: An equipment operator examines an exhauster on 15 August 1963. The original configuration could remove the gases at a rate of 100 pounds per second when the simulated altitude was 50,000 feet. Each of the thirteen 5,100-horsepower exhauster units contained two J33 compressor wheels that could be adapted to work in tandem or parallel. The number of units used could be varied to control the power load.[16] (NASA C–1963–65822)

Image 66: Each test chamber had a large primary cooler to reduce the temperature of the extremely hot engine exhaust before the air reached the exhauster equipment. The airflow exited the test section, passed through the chamber's 37-foot-long exhaust section, and then entered the cooler. PSL No. 2's primary cooler is seen here on 24 September 1952. (NASA C–1952–30771)

Image 67: Narrow fins or vanes inside the cooler were filled with water. As the 1,000°F to 2,000°F airflow passed between the vanes, heat was transferred from the air to the cooling water. The cooling water was cycled out of the system, carrying with it much of the exhaust heat. This photograph shows the vanes inside PSL No. 2's cooler on 20 January 1959. (NASA C–1959–49583)

Image 68: The airflow was then pumped through a secondary cooler, seen here on 8 October 1952. This secondary cooler cooled and scrubbed the exhaust gases from both test chambers to reduce explosion hazards before the gases were pumped through the exhausters in the Equipment Building and expelled into the atmosphere.[17] (NASA C–1952–30937)

Image 69: The cooling water for the primary and secondary coolers was circulated by the Circulating Water Pump House located across the street. The pump house, seen here in the foreground on 8 October 1952, contained the large water pumps and water-softening units. The water carrying the heat was sprayed down into the large, adjacent cooling tower. Roof fans in the wooden tower ceiling exhausted the hot air out as the water was diffused into the pools at the bottom. (NASA C–1952–30938)

Endnotes for Chapter 3

1. "PSL Model Now on Display in Cafeteria," *Wing Tips* (23 February 1951).
2. M. J. Krasnican, et al., "Propulsion Research for Hypersonic Flight," *NACA Lewis Flight Propulsion Laboratory Inspection October 7–10, 1957*. NASA Glenn History Collection, Inspections Collection, Cleveland, OH, p. 4.
3. E. W. Conrad and Martin Saari, Jr., "Full-Scale Engine Research, Part I—Research Techniques and Facilities," NACA Lewis Flight Propulsion Laboratory Inspection, 1954, NASA Glenn History Collection, Cleveland, OH.
4. Conrad and Saari, "Full-Scale Engine Research, Part I."
5. Conrad and Saari, "Full-Scale Engine Research, Part I."
6. Neal Wingenfeld interview, Cleveland, OH, 15 September 2008, NASA Glenn History Collection, Oral History Collection, Cleveland, OH.
7. Wingenfeld interview, 2008.
8. Wingenfeld interview, 2008.
9. Howard Wine interview, 4 September 2005, NASA Glenn History Collection, Oral History Collection, Cleveland, OH.
10. Wine interview, 2005.
11. "Major Research Facilities of the Lewis Flight Propulsion Laboratory, NACA Cleveland, Ohio, Engine Research Building," 17 July 1956, NASA Glenn History Collection, Cleveland, OH.
12. "Major Research Facilities of the Lewis Flight Propulsion Laboratory."
13. Wingenfeld interview, 2008.
14. Wingenfeld interview, 2008.
15. Robert Walker interview, Cleveland, OH, 2 August 2005, NASA Glenn History Collection, Oral History Collection, Cleveland, OH.
16. C. S. Moore, "Appendix A, Proposal for Propulsion Sciences Laboratory," NASA Glenn History Collection, Directors' Collection, Cleveland, OH, p. 7.
17. "Major Research Facilities of the Lewis Flight Propulsion Laboratory."

Image 70: An NACA Lewis researcher examines the massive General Electric J79 turbojet engine on 21 June 1956. One of the engine's applications was the Convair B-58 Hustler, the first operational supersonic jet bomber. (C–1956–42401)

· 4 ·
Cold War Weapons

Carlton Kemper, executive engineer of the NACA Aircraft Engine Research Laboratory (AERL) in Cleveland[c] boarded a DC-4 on 18 April 1945 and emerged in Europe at the dawn of the Cold War.[1] It was the eve of the bloody assault on Berlin, but the Allies were already fighting a new internal political battle, a battle for German technology.

The U.S. government initiated the secret Alsos Mission in 1943 to determine the extent of Germany's atomic weapon development program. In February 1945, the mission was expanded to assess advances in German technology of all types. With the fall of Berlin in sight, the United States and the Soviet Union raced to appropriate German expertise and intelligence for their own use.[2]

The effort, renamed Operation Paperclip, famously led to the capture and expatriation of German rocket scientists to the United States. What is less known is that a new team under the code name Operation LUSTY commenced work on 22 April 1945 to secure as much information and hardware as possible regarding German aircraft development for the U.S. Office of Scientific Research and Development.[3]

The operation included a contingent from the NACA, including Kemper from NACA's AERL and Russell Robinson, Henry Reid, and William Ebert from NACA's Langley Field. The mission proceeded in jeeps across the war-torn fields of France and into Germany in the wake of advancing Allied forces. The NACA team inspected secret laboratories, interrogated scientists, and seized troves of documents. Kemper concentrated his efforts on propulsion issues and engine test facilities. Another group collected aircraft and equipment for shipment back to the United States. At one laboratory, the NACA team found records hidden in a remote well.[4]

Two German aerospace advances were of particular importance: the axial-flow turbojet engine and the guided missile. The Germans had three types of turbojets, two rockets, and a pulse-jet aircraft in operation during the war.[5] The United States had acquired an early centrifugal turbojet from the British in 1942 and integrated it into a Bell airframe, but as a 1945 Alsos report stated, the United States was "very much behind the Nazis."[6]

Kemper was also impressed by German test facilities for turbojets. Several projects under construction would have surpassed the AERL's capabilities. Upon his return to the Cleveland lab in October 1945, Kemper warned, "Unless maximum use of our research equipment is made and unless we continue to expand our research facilities, when necessary, our present lead may be lost."[7]

The AERL began studying axial-flow and ramjet engines in 1945. Studies on rocket fuels and engine components began shortly thereafter. Although late

Image 71: Carlton Kemper was the AERL Executive Engineer when he left for six months on the Alsos Mission. The NACA capitalized on the wealth of information obtained during the mission, but it cost Kemper personally. Others, including Abe Silverstein and Addison Rothrock, stepped up during the transitional period, and Kemper's position became largely honorary. (NASA C–1943–01297)

to these fields, the lab's contributions helped U.S. development progress rapidly in the late 1940s. Yet, Kemper's words remained fresh in everyone's minds. The lab embarked upon a nearly 20-year period of almost constant expansion and development that had its roots at the very inception of the Cold War. The creation of the Propulsion Systems Laboratory (PSL) and its studies during the 1950s vividly reflect this situation.

The Nuclear Navaho

As World War II came to its final shuddering end in August 1945, the military created the requirements for a postwar missile program to deliver nuclear warheads. Industry was invited to bid on the program in late October. The following spring, the U.S. Air Force accepted North American Aviation's proposal to develop a reusable, winged, supersonic surface-to-surface long-range cruise missile. The weapon, known as Weapon System 104A, or the Navaho, added wings and improved rocket engines to the basic German V-2 rocket. The proposed 3,000-mile range was not sufficient for intercontinental travel, but it was more than adequate to hit Moscow from Britain or West Germany. The mysterious Navaho would prove to be a significant link between supersonic aircraft and rocketed space vehicles.[8]

The Navaho was a two-stage vehicle that relied on liquid-propellant rocket engines to boost it to altitude and two large ramjets to power it to its destination. Its unique navigation system allowed the missile to return to its base and land.

North American Aviation concentrated initially on the rocket engines for the booster. In his article, "The Navaho Inheritance,"[9] Russ Murray refers to North American's factory near Los Angeles as "America's Peenemunde."[d] He adds that North American's redesign of the liquid rocket propulsion system between 1948 and 1950 even superseded the German wartime development and provided the direction for the future progress on rocket engines for space. By 1950 this technology, along with the rapid advances in guidance and other subsytems, was integrated into the new Navaho missile.[10]

The program was broken into three phases with increasing ranges and propulsion systems: the 500-mile X-10 test vehicle powered by two Westinghouse J40 turbojets; the 1,500-mile G-26, or Navaho II, with two Wright XRJ47W5 ramjets; and the final 68-foot-long, 3,000-mile-range G-38 with two more powerful XRJ47-W-7 ramjets.[11]

Image 72: North American Aviation's Navaho Missile was an early attempt at a long-range cruise missile. (NASA C–1956–43440)

Navaho's 48-Inch-Diameter Powerplants

As work on the rocket engines for the booster progressed, North American contracted Curtiss-Wright's Wright Aeronautical Division in October 1950 to create the two powerful XRJ47 ramjets for the actual missile.

Ramjets were "just a big ugly tube that looked like an old afterburner," according to Bob Walker.[12] They are, indeed, among the simplest type of propulsion devices. Like other airbreathing engines, they were powered by combustion gases that were heated to high temperatures under pressure then exhausted. The ramjet's straightforwardness led to its efficiency. It was basically a tube with no moving parts. Fixed grated devices, referred to as "flameholders," produced a constant ignition source for the air passing through

---

[d] Peenemunde was the secret German rocket facility where Wernher von Braun and his associates created the V-1 and V-2 missiles in the early 1940s.

the ramjet. The downside was that ramjets could only be ignited at high speeds. This meant that they had to either be launched from an aircraft or with booster engines.[13]

The Navaho's 30,000 pounds of cruising thrust was provided by two 48-inch-diameter ramjets that *Flight* magazine claimed were "the most powerful airbreathing engine[s] yet made."[14] It is somewhat curious that Wright Aeronautical was contracted to develop these behemoths. Wright had been out of the nation's initial efforts in the early 1940s to develop turbojets so as to not inhibit the company's production of large reciprocating engines during World War II. After the war, Wright struggled to join the jet engine field with its J65 and compound engines but would give up its efforts by 1959.[15]

North American provided Wright with the specifications for the XRJ47 in July 1951. The engines were among the largest ever attempted, so it was not surprising that they were fraught with problems. Wright constructed an altitude test stand at its Wood-Ridge facility in 1949 specifically for ramjets, but its capability was limited.[16] In 1951 the military solicited the assistance of the NACA Lewis Flight Propulsion Laboratory in Cleveland and its brand new PSL facility. The project was classified 1-A, "An item designed to meet a potential threat against this nation, the lack of which would result in a national destruction and disaster in the event of war."[17]

PSL test engineers and mechanics began installing the XRJ47-W-5 for the G-26 in PSL No. 2 as the facility began its calibration tests in the spring of 1952. The Engine Research Division's massive, multifaceted XRJ47 test program formally began in October and lasted five years. The studies addressed specific XRJ47 performance issues, general ramjet concerns, and the advantages of different fuel types. The engine was tested in conditions that simulated the high-altitude cruising portion of its flight using both direct-connect and free-jet setups. The focus was on different elements of the combustion process.

Walker remembered being impressed, "It was designed to light at about [Mach] 1.6 or 1.8, something like that. And then it was designed to travel at Mach 3, which, we're talking 1950s here. It was pretty far out at that time."[18]

Image 73: *The Navaho X-10 vehicle shown in this 1954 photograph was flight powered by two Westinghouse turbojets. The larger G-26 and G-38 versions would use the Wright ramjets for flight. (NASA M-9142273)*

Pursuit of Power

Bob Walker

Walker was an ardent aircraft model maker while attending high school in the mid-1940s. Local aviation writer Charles Tracy had noted one of Walker's model competition victories in his weekly newspaper column.

Upon graduation in 1948, Walker was hired into Republic Steel's metrology lab. One of his classmates, however, was hired into the wood shop at the NACA Lewis lab. The classmate was getting paid to build the models, the very thing he enjoyed. When Walker heard about all the work being carried out on jet engines, he decided to send the NACA an application with a copy of Charles Tracy's article.

Lewis Director Ray Sharp appreciated the potential of young untrained model builders and had an established tradition of hiring them into the lab. Walker recalled, "I don't know whether that helped me or not, but at any rate, I got a chance to take the test and passed the test. I started in May of 1949, May the 16th, 1949."[19]

Image 74: Bob Walker. (NASA C–1953–34978)

Walker was immediately enrolled in Lewis's Apprentice Program. The school had been started in the early 1940s but was suspended during the war. Walker's 1949 class was the first since its reinstitution. The school taught a variety of trades, and the students were assigned 90-day details at various facilities.

"The object was to work in every facility where a mechanic would work on the whole lab. In addition to that, they included the instrumentation organization, so you learned how to make pressure rigs and thermocouples, things you'd be installing and taking apart. I got a real thrill, a broad education."[20]

Once Walker had proven himself, he was asked where he would prefer to be assigned. He jumped at the PSL. "It was the newest, hottest facility not only at the Lab, but probably in the world. It was ahead of all those other pressure tank units.... It seemed to me to be a good time to get in on the ground floor."[21]

Walker spent six months assisting in the initial calibration of the facility when PSL No. 2 Crew Chief Nick Ricciardi was drafted into the Korean War. Walker received his first promotion. "I wasn't able to take the job officially because I was still an apprentice, but they gave me the job anyway and didn't pay me the rate."[22] Walker continued as the PSL No. 2 Crew Chief after his apprentice class graduated in 1953. Ricciardi later returned to the division and enjoyed a lengthy career at the lab.

Walker worked his way up through the Test Installations Division in the 1960s. By 1971 he became head of the Electric Propulsion and Environment Test Branch and was part of the Advanced Test Satellite program.

By 1973 Walker transferred to the Technical Services Directorate, where his career would change dramatically. Over the next decade he oversaw Lewis's safety operations, equipment and supply, and the fabrication shop. Walker had risen to the division chief level by the time of his retirement in 1983.

Getting the Damn Engine to Ignite

The initial studies analyzed the combustion performance of a heavy-duty version of the XRJ47 using both the direct-connect and free-jet assemblies. The efficiency of variations on the combustor's internal geometry was investigated. The optimal combustor configuration was then installed on a flight-weight version of the engine and studied in a free jet by Henry Welna and Dwight Reilly. They found that the combustion efficiency matched what was found on the heavy engine and could be used as a baseline.[23]

Ramjets produced a tremendous amount of heat. Walker remembered, "It would throw out a pillar of blue flame, four feet in diameter at the beginning but spreading out as it went down the tank, just kind of fill that cooler with like a welding beam down through there."[24]

This heat threatened to burn up the entire engine, so NACA engineers developed a thin corrugated steel liner to prevent damage to the engine's external surfaces. The airflow between the two walls kept the engine shell from burning out.[25] "Occasionally, it didn't work," Walker added, "and we would be in there patching it up, rewelding the liner, changing the flameholder configurations for the fuel bars."[26]

Welna and Reilly analyzed three types of cooling liners or burner shells. They determined the optimal length, type of corrugation, and fasteners. Lewis engineers designed a new 0.030-inch-thick corrugated stainless steel liner with "hat" sections for extra support where the retaining rods held the shell to the engine.[27]

Welna and Reilly also examined three types of ignition using the existing flameholder. The ignition problem was a major concern. Walker explained, "Obviously, you throw this thing up into the air, one of the main things is you want the damn engine to ignite."[28] Welna determined that sparkplugs were useless during a previous XRJ47 test. They now tried electrically activated flares along the flameholder and the use of two special types of fuels.

"The one I hated the most was the flares," Walker remembered. The mechanics would have to climb through the cross-shaped temperature rake at the

Image 75: Damage to an XRJ47-W-5 combustion liner on 21 March 1955. (NASA C–1955–37979)

exhaust to install the flares inside the 48-inch-diameter engine. "[It was] kind of roomy in there, but you have the flameholder and then you have [a] box of squibs which were like flares with…little electrical exciters that set them off. You would have to wire these all on to the flameholder and bring the wires out and put it on the side." The test engineer would activate the flares to ignite the ramjet at the start of the test. "I always worried about one of those suckers going off while you were inside wiring it onto the flameholder," Walker reminisced. "You know, we didn't have much option then. It was all live; it was all a dream."[29]

Welna and Reilly tried different quantities and locations for the flares but concluded that the method was unreliable. The earlier tests on the heavy-duty XRJ47 had similar results. They then investigated the injection of boron triethyl and aluminum triethyl fuels into the combustor. Walker described loading a glass vial of one of the fuels into a stainless steel tube that ran through the chamber's view port and into the flameholder. "We loaded it in this tube like a rifle. At the appropriate time, we pulled a pin on it, and the vacuum would suck that glass tube down into it." The vial would smash against the flameholder and ignite the engine.[30] In the end, the injection of aluminum trimethyl into the flameholder's combustor proved to be the only consistent method of igniting the ramjet.[31]

A wide variety of different flameholder designs were studied by other ramjet researchers over the years. Two variations were analyzed on the XRJ47, an annular-piloted baffle-type flameholder and an annular can-type flameholder. Warren Rayle and Carl Wentworth compared the designs using three combustor lengths and presented their results graphically and tabularly.[32,33]

Welna teamed with Ivan Smith to study the XRJ47-W-5's fuel control system. The system used a control signal pressure to correct the flow during changes in altitude, speed, and angle of attack. An oscillograph recorded the step-changes for the fuel flow, inlet pressure, and fuel system pressure.[34]

Preliminary tests revealed that the XRJ47's side inlet produced flow separation and distorted flow profiles. Wright used a small-scale diffuser to develop some modifications to combat the distortions. Ivan Smith and John Farley tested the technique for the first time on a full-scale engine in the PSL during the fall of 1953 and early 1954. The modifications, which included a screen at the diffuser outlet and three arrangements of vortex generators in the diffuser duct, all improved efficiency. The greatest improvement, however, was obtained with the 12-stage vortex-generator configuration.[35]

Image 76: A view through the bellmouth cowl of the installation of the Wright 48-inch-diameter ramjet on 16 October 1952. The mechanic in the background has crawled through the cross-shaped temperature rake to work on the flameholder. (NASA C–1952–30950)

Image 77: Researcher Dwight Reilly examines the instrumentation on the XRJ47-W-5 in PSL No. 2 on 14 August 1953. He and Henry Welna were key researchers on the XRJ47 program. Reilly temporarily left Lewis in 1956 but returned the following year. In the early 1960s he managed testing and operations for the Space Nuclear Propulsion Office. In 1970 he was named Chief of the Space Power Facility.[36] (NASA C–1953–33390)

Image 78: Some of the different flameholder and combustor chamber designs for the Wright XRJ47-W-5 tested in the PSL between 1952 and 1955. The optimal efficiency for each configuration was determined. (NASA C–1954–35354, C–1952–31065, C–1954–36486, C–1952–31248, C–1955–39035, C–1952–31066, C–1954–36745, C–1952–31064, C–1955–40072, C–1954–35888, C–1956–41467, and C–1954–35889)

Taking Advantage of the Opportunity

Throughout the late 1940s and 1950s, the Engine Research Division was conducting general investigations into the altitude performance of ramjet engines. Previous tests had been performed on 16- and 28-inch-diameter ramjets in the Altitude Wind Tunnel and Four Burner Area. The researchers seized upon the opportunity presented by XRJ47 because of its unprecedented size and the fact that it was already installed in PSL No. 2.

One study sought to determine the most efficient method of combustion for long-range cruise missiles. It was determined that optimal fuel-air ratios are best maintained by confining the fuel injection to a specific portion of the combustor air. Henry Welna and Carl Meyer sought to test this procedure on the larger XRJ47. They analyzed the technique using two can-type flameholders and one baffle-type flameholder. The baffle configuration proved to be the most efficient.[37]

Warren Rayle, Ivan Smith, and Carl Wentworth modified the size of the exhaust nozzle of this combustor configuration to increase its efficiency from 90 to 100 percent and tested it in a free-jet setup. They analyzed performance using three combustor lengths and four fuel-distribution systems. They determined that the longer lengths produced the highest efficiency at lower fuel-to-air ratios, but that could be improved if the fuel was concentrated to the outer region of the burning stream.[38]

Image 79: Bob Walker examines a free-jet setup in PSL No. 2. Ferris Seashore and George Hurrell designed the Mach 1.75 installation specifically for the XRJ47 tests. Free jets provided simulated internal airflows for full-scale engines at a fraction of the cost of supersonic wind tunnels.[39]
(NASA C–1953–33387)

Image 80: A Wright 48-inch-diameter ramjet is installed in PSL No. 2 on 12 October 1952. A cast-iron mock-up of the inlet was used as researchers sought to establish a shock across the inlet to light the ramjet at supersonic speeds. (NASA C–1952–30961)

Performance-enhancing modifications to the ramjet combustor diffuser were developed at the NACA and elsewhere in the early 1950s. Welna and John Farley studied conical and reverse-bellmouth diffusers with variations of guide vanes, vortex generators, and splitter cones in the XRJ47 combustor inlet diffuser. The tests were run with a direct-connect configuration in PSL No. 2. Welna and Farley determined the optimal efficiencies for each modification and found that the combination of vortex generators and splitter cones was the most efficient.[40]

Researchers were also interested in studying elements that affected the control of the ramjet. This included diffuser shock movement and the recovery control performance range. Hurrell concentrated on controlling the ramjet's flight, first with a 16-inch-diameter ramjet in the 8- by 6-Foot Supersonic Wind Tunnel. Hurrell then studied indicial and frequency-response controls on the 48-inch-diameter ramjet. It was run in PSL No. 2 at simulated altitudes of 68,000 to 82,000 feet. Hurrell found that a fuel flow produced a varying period of dead time before the engine began its normal linear response.[41] In a separate study, George Vasu, Clint Hart, and William Dunbar also found dead time in the response.[42]

Lewis began exploring high-energy propellants such as liquid hydrogen, fluorine, and pentaborane in the early 1950s. Pentaborane had a 50-percent higher efficiency than JP-4. The researchers were looking to test the fuel on full-scale engines, and the XRJ47-W-5 ramjet that had been installed in PSL No. 2 for several years offered an excellent opportunity. During the two 1955 studies with different length combustors, the pentaborane produced 80- to 89-percent combustion efficiencies.[43]

Image 81: Relegated to serving as a museum exhibit, a Navaho missile and its booster are on display at the U.S. Space and Missile Museum in Florida. (Bob Walker)

Self-Immolation

While the XRJ47 tests were being performed in the PSL during 1953 and 1954, the Navaho X-10 with its two J40 turbojets was successfully flown from Cape Canaveral 27 times. It was the first turbojet to fly at twice the speed of sound.[44] Meanwhile, Wright began issuing the production version of the XRJ47 in 1956. A total of 59 XRJ47 engines were built in 1956 and 1957.[45]

Flight testing of the second Navaho phase, the G-26, began in late 1956 as the PSL tests were winding down. The first G-26 launch ended after only 26 seconds on 6 November. The next two launches yielded another booster failure and a massive launch pad explosion. The program was clearly in trouble. During the next attempt on 26 June 1957, the mission ended after only 4 minutes when the ramjets failed to ignite.[46]

The program was canceled almost immediately afterward, despite the fact that the initial G-38 vehicles were being manufactured. The successful test flight of an Atlas intercontinental missile in late 1956 had rendered the program superfluous, and the shift to intercontinental ballistic missiles would be accelerated after the space era was kick-started by the Sputnik launch in October 1957.[47]

The five remaining Navaho missiles in production were launched after the program was canceled. Two of these launches were exceptional: one traveled 500 nautical miles at Mach 3, and another, using only its autonavigator, traveled over 1,000 nautical miles.[48]

James Gibson summarized the program's legacy, "Navaho...was both a failure and a major triumph. The technology it developed actually made possible the vehicle that made the Navaho obsolete before [the] Navaho could be deployed."[49] The Navaho rocket system was used on the Redstone missile, its original propulsion system was used on Thor and Atlas boosters. These three missiles would be the backbone of the early space program. In addition, the Navaho guidance system was used by the USS Nautilus nuclear submarine on its groundbreaking journey underneath the polar ice cap.[50]

Coda

Before the program's cancellation, the air force requested that Lewis undertake similar studies for the Wright XRJ47 W-7 ramjets that were to power the final G-38 Navaho. The studies would determine the effect of high Mach numbers on engine structure, burner performance, shock positioning controls, and component efficiencies.[51] North American officials were also interested in using Lewis's supersonic wind tunnels for inlet

Image 82: Gene Wasielewski. (NASA C-1955-39823)

and internal flow studies.⁵² Neither of these programs was carried out, however.

It is also interesting to note that Curtiss-Wright was so impressed with PSL's performance during the XRJ47 tests, that it sought to build a similar facility at its new 85-square-mile site at Wood-Ridge, New Jersey. Curtiss-Wright decided to hire an expert, Gene Wasielewski. Wasielewski left Lewis in 1956 to take on the $7.7 million project. Like the PSL, the facility contained two altitude test cells for testing large engines. Curtiss-Wright claimed that the facility, which could simulate speeds of 3,500 miles per hour and an altitude of 95,000 feet, was the largest privately operated supersonic ramjet laboratory in the nation.⁵³

Defense Missiles

Missiles were not only being developed to attack but also defend. The Bomarc was the first U.S. long-range interceptor missile for combating Soviet bombers. In 1949 the air force contracted with Boeing to research the possibility of a supersonic anti-aircraft missile. The University of Michigan Aeronautical Research Center was soon brought in as a partner, and the program was named Bomarc. Development officially began in January 1951, and the first test flight was on 10 September 1952, just as the PSL was being completed.⁵⁴

The Bomarc was launched vertically using an Aerojet rocket engine. Its final climb to 60,000 feet and cruising speed of Mach 2.5 were powered by two 28-inch-diameter Marquardt RJ43-MA-3 ramjet engines.

Marquardt was a California-based company that had focused exclusively on ramjets since 1944. Its Ogden, Utah, plant produced over 1,200 ramjets for the Bomarc program.⁵⁵

Marquardt had its own test facilities at Van Nuyes, California, including altitude test cells, but the Bomarc program required outside consultation and testing.

Image 83: Squadron of Bomarc missiles in the 1960s, each with its two 24-inch-diameter Marquardt ramjets visible. Ten missile sites with a total of 500 weapons were deployed around the perimeter of the United States, including two Canadian sites. (U.S. Air Force)

Image 84: A 28-inch-diameter Marquardt RJ43 ramjet engine being readied by two technicians on 5 October 1954 for a free-jet test in PSL No. 1. It was tested at Mach 2.5 and at altitudes up to 65,000 feet. (NASA C–1954–36898)

Their engineers studied Mach number tables from the U.S. Navy's Project Bumblebee, university information on flame stabilization and spread during combustion, and performance curves and charts generated by the NACA.[56]

Lewis's Engine Research Division took on a broad study of the RJ43's altitude performance throughout 1954 and 1955. The small size of the engine made it suitable for combustion studies in the 10-foot-diameter cells in the Four Burner Area. Carl Wentworth, William Dunbar, and Robert Crowl, however, used PSL No. 1 to study the engine's shock-positioning control unit and dynamic response. The pneumatic shock-positioning control unit was used to keep the ramjet on the correct flight path.[57] The researchers also examined and plotted the dynamic response of the engine at varying speeds and altitudes.[58] The PSL test data were then turned over to Marquardt to verify their design.

The U.S. Air Force originally intended to deploy 6,300 Bomarc missiles at over 50 launch sites. This number of sites was reduced to 40 by the time of the first production unit flight in 1955. The missile flew 250 miles and successfully hit its target. The number of potential sites was further reduced to 22 in 1956. In a 2 October 1957 test firing, a Bomarc missile was able to intercept one of the turbojet-powered Navaho missiles travelling in excess of Mach 2.[59]

Fiscal disagreements continued, however, as new missile types became available in the late 1950s. When final deployment was completed in 1962, there were only 10 Bomarc sites.[60] The program was canceled by Congress in 1970, and the last missile was retired two years later. Despite the political infighting, Boeing emerged from its first missile program as experts in large-scale systems integration. They would soon be involved with the Minuteman Missile program.[61]

Howard Wine

Wine's four years as a flight engineer in the air force gave him an advantage as he was hired at NACA Lewis. Wine had been out of the service nearly a year when his sister-in-law, a Lewis chemist, encouraged Wine to apply at the lab. He was one of only three journeymen when he was hired in January 1954. This meant that Wine, unlike his contemporaries, could forgo the standard apprentice training.[62]

After his midday interview, Wine was introduced to his new supervisor at the PSL, Paul Rennick. Wine recalled Rennick telling him, "Since you are already here, why don't you stay until 4:30…and then tomorrow, start second shift." The crews rotated between first and second shifts every other week, while the overnight crew remained on third shift. Wine had been unaware that Lewis's facilities operated at night, but he overcame his reluctance for shift work and formally began his 33-year career the next day.[63]

Wine always had an eye on advancement and regularly petitioned PSL Section Head Bud Meilander for one of the already-filled crew chief positions. At the time, the research engineers were increasingly balking at coming in on third shift to run the PSL tests, so a new protocol was instituted. Meilander called a meeting to inform the staff that the crew chiefs would assume the engineers' responsibility for running the tests. Wine recalled chasing Meilander out the door after the meeting. "Gee, Bud, I've been sitting in with the engineers. I know how to go ahead and do this." Meilander cautioned Wine, "Well, slow down, Howard. We've got to give the crew chief a chance, but you can sit in with them and kind of help them along."[64]

Wine decided to pursue an engineering degree at Baldwin-Wallace College, which had a long relationship with Lewis. He switched to third shift and took morning classes for two years. By the late 1950s he was appointed as a crew chief at the PSL, however, and was running out of available morning courses at school. He remembered frankly asking the new division chief, Bill Egan, whether to finish his engineering degree or continue his career in the technical field. Egan replied,

Image 85: Howard Wine. (NASA C–1965–01796)

"Well, Howard, I tell you, we think you've got a lot of potential here in the area that you are in, and that you know you are making as much money now as a starting engineer. So if you are looking at the money progression, I think you would be better off just staying right where you are at."[65]

Wine forewent his degree, and returned to his crew chief post at the PSL. "PSL 1 and 2 by now I knew backwards and forwards," he recalled, "just loved coming up and punching buttons and climbing around, blowing up engines and things like that."[66]

In the early 1960s, Meilander was promoted and Wine was selected to take his place as section head. Rennick and Carl Betz remained as PSL shift supervisors. Wine remembered, "I could imagine them thinking, 'This young punk coming in here and he's going to be our boss,' but any rate we had [a] great [relationship]."[67] Wine went on to work his way up to assistant division chief and later served as an assistant to the director. Wine retired in 1987.

Testing the Turbojets

The PSL was built to test jet engines. Despite periodic ramjet and rocket studies, the PSL was primarily used to test jet engines throughout most of its operational life. Yet, the turbojet research in the 1950s seemed a bit bland compared with the emerging missile and rocket fields. Bob Walker recalled that PSL No. 2 tested "rockets and weird stuff" in the 1950s, while PSL No. 1 performed more conventional turbojet studies. "We [the missile people] always considered their [jet engine] work, kind of, you know…[laughs and gestures]. It was funny in those days, everything was competitive."[68]

The first American turbojets had been built by General Electric (GE) and Westinghouse in 1942. NACA's Cleveland lab performed an immense amount of work in developing these engines and their successors during World War II and in the immediate aftermath. At the request of the military and engine manufacturers, the Altitude Wind Tunnel was used to analyze almost every turbojet design of the period. As the development of the turbojet progressed, so did the test requirements. The PSL was designed specifically to handle the larger jet engines of the 1950s and 1960s.

Full-scale aircraft engine testing had been a primary strength of the lab since its inception. Engine components such as the compressor, turbine, combustor, and exhaust nozzle may operate flawlessly when isolated, only to malfunction when integrated into an engine system. For example, distorted airflow from the compressor may hamper the combustor's operation. Other complete engine system problems included combustor blowout at high altitudes, component overheating, and unstable control systems.

General Electric 1950s Successes

GE's Schenectady, New York, plant began developing a series of axial-flow engines in the mid-1940s, including the 5,800-pound-thrust J47. The J47 became a tremendous success. It was used to power the successful F-86 Sabre fighter and later became the first jet engine used in a commercial aircraft.[69]

GE soon developed the 9,500-pound-thrust successor, the J73, for the air force's F-86H Sabrejet. The F-86H was a low-altitude bomber version of the F-86 fighter. The J73 was the first engine tested in PSL No. 1. Between 1952 and 1954 the performance

Image 86: Martin Saari points to a GE J79 turbojet engine model at the PSL during an NACA Inspection talk on 8 October 1957. The talk is about full-scale engine testing in Lewis's large facilities. (NASA C-1957-46136)

Pursuit of Power

Image 87: GE J73 compressor stator blades seen on 10 November 1953. (NASA C–1953–34173)

characteristics of the J73-GE-1A (the first in a series of J73 production engines) were studied extensively in the PSL over a range of altitudes and speeds.

Lewis researchers tracked the engine performance characteristics and combustion efficiency.[70] Other Lewis researchers examined the range of combustion and compressor efficiency using a 10-percent larger turbine nozzle to avoid compressor surge.[71]

Another J73 model, the YJ73-GE-3, was run almost 200 times in the PSL using a normal turbine nozzle. The overall engine performance, as well as that of its various components, were studied and graphically mapped.[72]

Although the J73 was more powerful than its predecessor, the limitations of the F-86H airframe prevented the engine from performing at its optimal capacity. GE was also having difficulty producing enough of the engines and spare parts to satisfy the F-86H program. Although the bomber version of the F-86 was canceled by 1956, the J47-powered F-86 fighter set a world's speed record at the September 1948 National Aircraft Show in Dayton and played a key role in the Korean War.[73]

GE's next-generation turbojet was the Collier-Trophy-winning 11,900-pound-thrust J79.[e] It was designed with customized variable stator vanes to process the massive airflow encountered during supersonic flight. The variable stator vane concept allowed new fighter jets to fly at twice the speed of sound for the first time.[74]

The U.S. Air Force requested that Lewis improve the J79's afterburner performance. PSL testing demonstrated that modifications to the fuel system and flameholder increased the combustion efficiency and reduced the pressure drop.[75] Researchers also collected performance data from a J79's independently controlled variable ejector assembly.[76]

The J79 was a significant accomplishment for GE. It powered key Vietnam-era aircraft such as the F-4 Phantom, the F-104 Starfighter, the B-58 Hustler, and a number of civil transport aircraft.[77]

Image 88: F-4 Phantom powered by two J79 turbojets, 2005. (U.S. Air Force)

Ill-Fated Avro Project

Canada's Avro Aircraft Limited began design work on the CF-105 Arrow jet fighter in 1952. The company's Orenda branch suggested building a titanium-based PS.13 Iroquois engine following development problems with the British engines that Avro had originally intended to use. The 10-stage, 20,000-pound-thrust Iroquois would prove to be more powerful than any contemporary U.S. or British turbojet. It was also significantly lighter and more fuel efficient.[78]

---

[e] The Robert J. Collier Trophy is awarded annually "for the greatest achievement in aeronautics or astronautics in America, with respect to improving the performance, efficiency, and safety of air or space vehicles, the value of which has been thoroughly demonstrated by actual use during the preceding year."

The new Avro aircraft and engine were developed simultaneously. The design began in March 1954 and proceeded rapidly. The Iroquois engine was ground tested thousands of times between its first run in December 1954 and 1958. This testing included studies at the new Arnold Engineering Development Center, flight tests under a B-47, inlet studies in Lewis's 8- by 6-Foot Supersonic Wind Tunnel, and 141 hours of investigations at the PSL.[79]

The air force asked Lewis to study the basic performance of an Iroquois engine. An Iroquois was sent to Cleveland in April 1957. John Groesbeck, John McAulay, and Daniel Peters of the Engine Research Division were the primary researchers for the Lewis testing. Early studies demonstrated severe compressor stall at high altitudes.

Initial PSL studies determined the Iroquois's windmilling and ignition characteristics at high altitude. The tests were run over a wide range of speeds and altitudes with variations in exhaust-nozzle area. After operating for 64 minutes, the engine was reignited at altitudes up to the 63,000-foot limit of the facility. The researchers found that decreasing the nozzle area reduced windmilling.[80] Peters and Groesbeck also made a brief study of two afterburner configurations with different fuel-injection patterns.[81]

The manufacturer modified two engines by adding a set of variable guide vanes at the high-pressure compressor inlet and minor alterations to the compressor. McAulay and Groesbeck studied the operating limits of the original engine in the PSL to better understand the problem. They then studied the two modified engines. They found that severe radial flow distortions at the compressor inlet reduced the high-pressure compressor stall limit. Various modifications were studied that reduced the occurrence of stall but did not totally eradicate the problem.[82]

The Arrow jet fighter made its initial flight in March 1958 powered by a substitute engine. In February 1959, however, both the engine and the aircraft programs were canceled. As with the Navaho and many other military programs of the mid-1950s, the transition of the superpowers' weaponry to ballistic missiles had rendered the Avro Arrow prematurely obsolete.[83] The Iroquois had been run for over 7,200 hours, but it had never been integrated into the Arrow aircraft and flown.[84]

The cancellation left over 200 Canadian aerospace engineers out of work. Twenty-five of them immediately joined the new U.S. space agency. Many others joined later or took positions with large U.S. contractors working on the human space program.[85]

Image 89: Brake testing of the Avro Arrow fighter with substitute engines. The aircraft, without the Iroquois engine, was tested extensively in the mid-1950s, including on missile boosters launched from the NACA Langley Aeronautical Laboratory. (Canadian Department of National Defense)

Pursuit of Power

Image 90: John McAulay (to the right) and a technician install the Orenda Iroquois engine on 29 April 1957. (NASA C–1957–45206)

The Avro story provides an excellent example of the state of the field of aviation at a critical transitional point. Turbojet and ramjet engine capabilities had grown exponentially over the past decade, and in the Iroquois case, had become more efficient. The civilian airline industry would soon be adding jet engines to its fleets. Yet, in many ways the success was overshadowed by the emergence of rocket systems in the mid-1950s. This new technology would dominate both Lewis and the PSL for the next 10 years.

Endnotes for Chapter 4

1. "Kemper Reports on Foreign Labs," *Wing Tips* (2 November 1945).
2. Sterling Michael Pavelec, *The Jet Race and the Second World War* (Westport, CT; Praeger Security International, 2007).
3. "Operation LUSTY," National Museum of the United States Air Force, http://www.nationalmuseum.af.mil/factsheets/factsheet.asp?id=1608 (accessed 27 April 2011).
4. "Kemper Reports on Foreign Labs."
5. Pavelec, *The Jet Race and the Second World War.*
6. Alsos Mission Report, 4 March 1944, National Archives, Intelligence Division, Alsos Mission File 1944–1945, Entry 187, Box 137, as referenced by Virginia Dawson, *Engines and Innovation: Lewis Laboratory and American Propulsion Technology* (Washington, DC: NASA SP–4306, 1991).
7. "Kemper Reports on Foreign Labs."
8. Russ Murray, "The Navaho Inheritance," *American Aviation Historical Society*, 19, no. 1 (Spring 1974): 17.
9. Murray, "The Navaho Inheritance," 17.
10. Murray, "The Navaho Inheritance," 18.
11. James N. Gibson, *The Navajo Missile Project* (Atglen, PA: Schiffer Publishing Ltd., 1996).
12. Robert Walker interview, Cleveland, OH, 2 August 2005, NASA Glenn History Collection, Oral History Collection, Cleveland, OH.
13. Abe Silverstein, "Ramjet Propulsion," c. later 1940s, NASA Glenn History Collection, Cleveland, OH.
14. "Guided Missiles 1956," *Flight*, (7 December 1956).
15. "Aero Engines 1959, The Propulsion Spectrum," *Flight* (20 March 1959).
16. "History and Organization of Wright Aeronautical Division Curtiss-Wright Corporation," http://www.scribd.com/doc/12467677/History-of-Curtis-Wright-Aeronautical-Company

17. Harold Robbins to Abe Silverstein, "Priority of Power Plant Laboratory Projects," 28 January 1951, NASA Glenn History Collection, Directors' Collection, Cleveland, OH.
18. Walker interview, 2005.
19. Walker interview, 2005.
20. Walker interview, 2005.
21. Walker interview, 2005.
22. Walker interview, 2005.
23. Henry Welna and Dwight Reilly, *Preliminary Evaluation of Flight-Weight XRJ47-W-5 Ram-Jet Engine at a Mach Number of 2.75* (Washington, DC: NACA RM E55G22, 1954).
24. Walker interview, 2005.
25. Welna and Reilly, *Preliminary Evaluation of Flight-Weight XRJ47-W-5 Ram-Jet Engine.*
26. Welna and Reilly, *Preliminary Evaluation of Flight-Weight XRJ47-W-5 Ram-Jet Engine.*
27. Welna and Reilly, *Preliminary Evaluation of Flight-Weight XRJ47-W-5 Ram-Jet Engine.*
28. Walker interview, 2005.
29. Walker interview, 2005.
30. Walker interview, 2005.
31. Welna and Reilly, *Preliminary Evaluation of Flight-Weight XRJ47-W-5 Ram-Jet Engine.*
32. Warren Rayle, Ivan Smith, and Carl Wentworth, *Preliminary Results From Free-Jet Tests of a 48-Inch Diameter Ramjet Combustor With an Annular-Piloted Baffle-Type Flameholder* (Washington, DC: NACA E54K15, 1955).
33. Carl Wentworth, Wilbur Dobson, and Warren Rayle, *Preliminary Results From Free-Jet Tests of a 48-Inch Diameter Ramjet Combustor With an Annular Can-Type Flameholder* (Washington, DC: NACA RM E54L07, 1955).
34. Henry Welna and Ivan Smith, *Preliminary Transient Performance Data on the Fuel Control of the XRJ47-W-5 Ramjet Engine* (Washington, DC: NACA RM E54A25, 1954).
35. John Farley and Ivan Smith, *Preliminary Performance Data Obtained in a Full-Scale Free-Jet Investigation of a Side-Inlet Supersonic Diffuser* (Washington, DC: NACA RM SE54J22, 1954).
36. "Reilly Appointed SPF Chief," *Lewis News* (14 August 1970).
37. Carl Meyer and Henry Welna, *Investigation of Three Low-Temperature-Ratio Combustor Configurations in a 48-Inch Diameter Ramjet Engine* (Washington, DC: NACA RM E53K20, 1954).
38. Rayle, Smith, and Wentworth, *Preliminary Results… Annular Piloted Baffle-Type Flameholder.*
39. Ferris Seashore and Herbert G. Hurrell, *Starting and Performance Characteristics of a Large Asymmetric Supersonic Free-Jet Facility* (Washington, DC: NACA RM–E54A19, 1954).
40. John Farley and Henry Welna, *Investigation of Conical Subsonic Diffusers for Ramjet Engines* (Washington, DC: NACA RM–E53L15, 1954).
41. Herbert Hurrell, *Experimental Investigation of Dynamic Relations in a 48-Inch Ramjet Engine* (Washington, DC: NACA RM E56F28, 1957).
42. George Vasu, Clint Hart, and William Dunbar, *Preliminary Report on Experimental Investigation of Engine Dynamics and Controls for a 48-Inch Ramjet Engine* (Washington, DC: NACA RM E55J12, 1956).
43. Warren Rayle, Dwight Reilly, and John Farley, *Performance and Operational Characteristics of Pentaborane Fuel in 48-Inch Diameter Ramjet Engine* (Washington, DC: NACA RM E55K28, 1957).
44. Murray, "The Navaho Inheritance," 20.
45. Aircraft Engine Historical Society, "Summary Curtiss-Wright Engine Shipments All Plants, All Models, All Licensees Except Curtiss Commercial Production Prior To 1931 Not Included" (Transcribed from a Curtiss-Wright document of the early 1960s), http://www.enginehistory.org/Wright/WrightProd.pdf (accessed 10 May 2011).
46. Ed Kyle, "Pilotless Bomber: US Space Technology Incubator," *Space Launch Report* (2005), http://www.spacelaunchreport.com/navaho1.html (accessed 10 April 2011).
47. Jack Raymond, "Navaho Missile Being Discarded," *New York Times* (12 July 1957).
48. Murray, "The Navaho Inheritance," 18.
49. James Gibson, "The Navaho Project—A Look Back," *North American Aviation Retirees Bulletin* (Summer 2007).
50. Murray, "The Navaho Inheritance," 18.
51. Trygve Blom, "Propulsion Projects Allocation and Priority Group," 7 April 1955, NASA Glenn History Collection, Test Facilities Collection, Cleveland, OH.
52. Smith deFrance to NACA Lewis, "Possibility of Expediting the Inlet Development of Several Military Aircraft by the Transfer of Some Investigations From the Ames Unitary Plan Wind Tunnel to Lewis Wind Tunnels," 27 December 1955, NASA Glenn History Collection, Directors' Collection, Cleveland, OH.
53. "History and Organization of Wright Aeronautical Division Curtiss-Wright Corporation."
54. Michael Lombardi, "Reach for the Sky: How the Bomarc Missile Set the Stage for Boeing to Demonstrate Its Talent in Systems Integration," *Boeing Frontiers* (2 June 2007).

55. Carl R. Stechman and Robert C. Allen, *History of Ramjet Propulsion Development at the Marquardt Company—1944 to 1970* (Reston, VA, AIAA–2005–3538, 2005), p. 2.
56. Stechman and Allen, *History of Ramjet Propulsion Development at the Marquardt Company.*
57. R. Crowl, W. R. Dunbar, and C. Wentworth, *Experimental Investigation of a Marquardt Shock-Positioning Control Unit on a 28-Inch Ram-Jet Engine* (Washington, DC: NACA RM–E56E09, 1956).
58. Carl Wentworth, William R. Dunbar, and Robert J. Crowley, *Altitude Free-Jet Investigation of Dynamics of a 28-Inch Diameter Ramjet Engine* (Washington, DC: NACA RM–E56F28b, 1957).
59. Richard P. McMullen, *History of Air Defense Weapons, 1946–1962*, Air Defense Command Historical Study Number 14 (Historical Division, Office of Information, HQ Air Defense Command, 1963), chap. 6.
60. McMullen, *History of Air Defense Weapons, 1946–1962.*
61. Lombardi, "Reach for the Sky."
62. Howard Wine interview, Cleveland, OH, 4 September 2005, NASA Glenn History Collection, Oral History Collection, Cleveland, OH.
63. Wine interview, 2005.
64. Wine interview, 2005.
65. Wine interview, 2005.
66. Wine interview 2005.
67. Wine interview, 2005.
68. Walker interview, 2005.
69. "The History of Aircraft Engines," GE Aviation (2010), http://www.geae.com/aboutgeae/history.html (accessed 10 December 2010).
70. Carl E. Campbell and E. William Conrad, *Altitude Performance Characteristics of the J73-GE-1A Turbojet Engine* (Washington, DC: NACA RM–E53I25, 1954).
71. Carl E. Campbell and Adam Sobolewski, *Altitude-Chamber Investigation of J73-GE-1A Turbojet Engine Component Performance* (Washington, DC, NACA RM–E53I08, 1954).
72. Harold R. Kaufman and Wilbur F. Dobson, *Performance of YJ73-GE-3 Turbojet Engine in an Altitude Test Chamber* (Washington, DC: NACA RM E54F22, 1955).
73. "North American F-86 (Day-Fighter A, E and F Models)," National Museum of the United States Air Force (2011), http://www.nationalmuseum.af.mil/factsheets/factsheet.asp?id=2297 (accessed 10 December 2010).
74. Harry E. Bloomer and Carl E. Campbell, *Experimental Investigation of Several Afterburner Configurations on a J79 Turbojet Engine* (Washington, DC: NACA RM–E57I18, 1957).
75. Bloomer and Campbell, *Experimental Investigation of Several Afterburner Configurations.*
76. William Greathouse and Harry Bloomer, *Preliminary Internal Performance Data for a Variable-Ejector Assembly on the XJ79-GE-1 Turbojet Engine, I—Nonafterburning Configurations* (Washington, DC: NACA RM–56E23, 1956).
77. "The History of Aircraft Engines."
78. John McAulay and Donald Groesbeck, *Investigation of a Prototype Iroquois Turbojet Engine in an Altitude Test Chamber* (Washington, DC: NACA RM SE58E26, 1958).
79. "Aero Engines 1959."
80. Daniel Peters and John McAulay, *Some Altitude Operational Characteristics of a Prototype Iroquois Turbojet Engine* (Washington, DC: NACA RM–SE58F17, 1958).
81. Donald Groesbeck and Daniel Peters, *Altitude Performance of the Afterburner on the Iroquois Turbojet Engine* (Washington, DC: NACA RM–SE58G01, 1958).
82. McAulay and Groesbeck, *Investigation of a Prototype Iroquois Turbojet Engine.*
83. Emmanuel Gustin, "Interceptor Rex—the Avro CF-105 Arrow" (1996), http://www.avro-arrow.org/Arrow/written_history.html (accessed 10 December 2010).
84. "Aero Engines 1959."
85. Chris Gainor, *Arrows to the Moon: Avro's Engineers and the Space Race* (Burlington, Ontario: Collector's Guide Publishing Inc., 2001).

Image 91: John Kobak and an engineer examine a drawing and model of the Atlas/Centaur while a technician inspects the setup of a Pratt & Whitney RL-10 rocket engine in a PSL test chamber on 13 November 1962. Two of the 15,000-pound-thrust engines were used to power the Centaur second-stage rocket. (NASA C–1962–62465)

· 5 ·
The Rocket Era

A still quietness settled as the laboratory's strangely shaped shadows grew longer. Evening was falling, and most of the staff had already left the campus when the lab's squad cars rolled up to the Propulsion Systems Laboratory (PSL). The officers purposely placed the wooden roadblocks in the streets surrounding the facility. They then methodically walked the hallways of the nearby buildings evicting anyone still at their desk.

Those inside the PSL, however, remained. "We were locked up in the control room," Neal Wingenfeld recalled, "We were not allowed to leave…They would barricade the roads and nobody was allowed in or out. We were in here, so we were goners if anything happened."[1]

The technicians and test engineers were hunkered in an unpressurized control room, just feet away from a blazing rocket engine and hundreds of gallons of liquid hydrogen and liquid oxygen.

John Kobak recalled working all day and returning in the evening. "I'd go home and eat and come back. As soon as people go home, we'd bring all that stuff in. And about seven o'clock we'd lock down and run until about eleven o'clock…or however long it took. We'd do that at least two or three times a week. It was fun."[2]

It was early 1961, and the first liquid-hydrogen rocket engine, the Pratt & Whitney RL-10, was under investigation in PSL No. 1. Two RL-10s would power the Centaur second-stage rocket, a key element of the early Apollo Program. These early PSL tests were critical to resolving many of the engine's operating problems and led to NASA Lewis Research Center's[c] greater involvement in the space program.

Cleveland Rockets

The Cleveland lab's interest with rockets had started 15 years before. By the mid-1940s a small group of researchers in the Fuels and Lubrication Division had begun working with a variety of high-energy propellants, oxidizers, and small rocket engines. A series of cinderblock test cells, referred to as the "Rocket Lab," were built behind protective earthen mounds in a far corner of the Cleveland lab. The group's efforts began producing significant results in the early 1950s, and Abe Silverstein promoted the group to a division branch. Design work began in 1952 on a larger dedicated static rocket stand, the Rocket Engine Test Facility. (For a complete history of the lab, see Dawson.[3])

The military, the Jet Propulsion Laboratory, universities, and commercial firms such as Aerojet were feverishly developing rocket engines and experimenting with propellants in the 1940s, including the cryogenic liquid hydrogen. By 1950 the military became frustrated with the difficulties associated with liquid hydrogen and cut research funding. The NACA Lewis Flight Propulsion Laboratory was also interested in high-energy propellants and took up the cryogenic mantle. For more indepth histories of Glenn's liquid-hydrogen work, see Dawson and Sloop.[4,5])

By 1955 Lewis researchers had demonstrated that a combination of liquid hydrogen and liquid oxygen was the most efficient propellant and were confident that performance problems could be resolved.[6] Silverstein immediately became an enthusiastic advocate. At the air force's request, the lab undertook a one-year crash program to demonstrate that liquid hydrogen could be used for aircraft engines. The project yielded several liquid-hydrogen-powered flights of a B-57B Canberra over Lake Erie in early 1957.

Image 92: The PSL, partially obscured by a cloud of steam, was located in a densely populated area of NACA's Cleveland lab. The 8- by 6-Foot Supersonic Wind Tunnel is in the foreground, the Chemistry Lab, the PSL Operations Building, and Instrument Lab are in the background. Other nearby facilities, including the Engine Propeller Research Building and the Engine Components Research Laboratory, are not visible. (NASA C–1955–38660)

Meanwhile, the NACA's Triennial Inspection was to be hosted by Lewis in October 1957. Over 2,000 military, scientific, and industry representatives visited during the four-day open house. Lewis was anxious to demonstrate its rocket engine and fuels work, but NACA Secretary John Victory had ordered them to downplay those efforts. Congressman Albert Thomas's crusade was still fresh in Victory's mind, and he feared recriminations from the congressional and military visitors who might think the NACA should be concentrating on aeronautics and leave rocket weaponry to the military.[7]

On 4 October, however, just days before the Inspection opened, the Soviet Union sent Sputnik I sailing around the Earth. Public and congressional outcry over the nation's weak space program changed everything. The rocket talks were reinserted into the Inspection agenda, and the event was a watershed moment in Lewis's history.[8]

Full-scale engine research veterans William Fleming and Martin Saari gave presentations at the PSL on recent propulsion research for hypersonic flight and full-scale engine testing. Speakers at other stops described high-energy fuels and rocket component research.[9] The Lewis Inspection proved to be a bright spot for the nation's aerospace engineers in the Sputnik aftermath. Not only was rocket research acceptable, it was demanded. Lewis, including the PSL, quickly refocused almost its entire research effort toward space issues.

Transformation

According to Howard Wine, Sputnik had an almost immediate effect on the PSL. "They were running the Iroquois engine…and on the night that the Sputnik went up, we were scheduled to run that again. And [the supervisor] said, 'Hold that off. We're going to change over and get into rocket testing.' So we didn't even run that night. That very night we just changed it."[10]

Image 93: In the late 1950s the PSL turned its attention away from jet engines and ramjets to small rockets. Many small rockets and rocket components were tested in the PSL between 1957 and 1963.[11] This 30 September 1958 photograph shows a technician installing an isentropic rocket nozzle in PSL No. 2. (NASA C–1958–48831)

NACA records indicate that the last run of an airbreathing engine in the PSL was on 8 March 1958, but there is little doubt that planning for a long-term study of rocket nozzles in simulated altitudes began the previous October. The newly formed Propulsion Systems Division took over most of the PSL research. The changeover from airbreathing engines to rockets in 1958 resulted in the PSL's lowest annual usage rate to date, but that would soon change.[12]

NASA was officially established on 1 October 1958. NASA expanded from NACA's handful of research laboratories to a dozen centers specializing in spaceflight. The NACA Lewis Flight Propulsion Laboratory became the NASA Lewis Research Center. The staff and facilities at all of the centers expanded. In addition, NASA moved into the management of large projects and the oversight of work performed by industry contractors.

After leading the center for 20 years, Ray Sharp retired in December 1960. He passed away seven months later. In October 1961 Abe Silverstein, Sharp's trusted partner, returned from a three-year detail at NASA Headquarters to assume Sharp's former position of Center Director.

Hypersonic Heating

Lewis researchers were frustrated with their inability to simulate normal atmosphere at high temperatures in their facilities. As early as 1956, Lewis engineers sought ways to increase combustion air to hypersonic temperatures.[13]

Lewis's increased efforts to enable supersonic and hypersonic speeds for missiles and space travel meant that many of its existing facilities needed improvement. Nineteen facilities were built or reassigned for space-related testing during the Apollo Program.[14]

PSL modifications included cryogenic fuel-pumping capabilities, an upgraded control room, and the installation of a pebble bed heater in PSL No. 2.

The pebble bed heater simulated the high temperatures produced at supersonic and hypersonic speeds by creating 3,500°F airflows through a 24-inch-diameter test section. The heater was a cylindrical brick structure filled with 10 tons of aluminum-oxide pebbles that stood vertically beneath PSL No. 2. A gas-fired heater initially brought the bed up to proper temperature. The flame was then closed, and cool air was passed through the bed. The hot pebbles warmed the airflow as it passed through the bed. The heated air expanded through a nozzle into the test section. This resulted in a small hypersonic wind tunnel inside the PSL chamber.[15] This hypersonic tunnel was used to study cooling and dissociation on 16-inch-diameter ramjet engines for durations from 3 to 5 minutes.[16]

The temperamental device required a couple of hours to generate enough heat for a run of several minutes. Walker explained, "It was kind of hectic thing to run, and you were never quite sure how it was going to run."[17]

A thrust rig was built in PSL No. 2 to complement the pebble bed heater. Researchers used the rig to study nozzle configurations for thrust vectoring tests in 1961.[18]

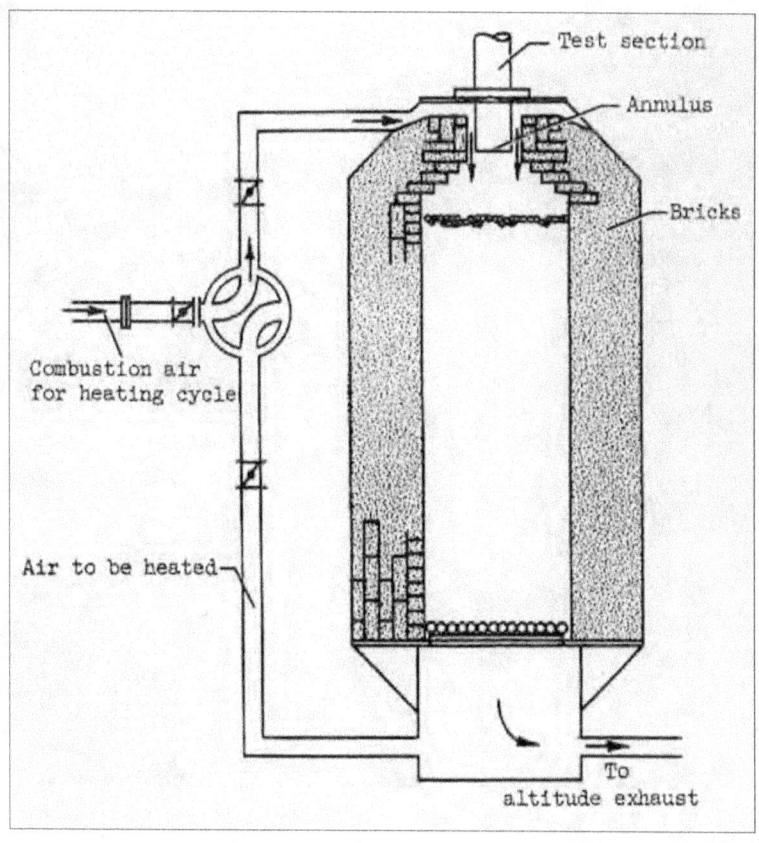

Image 94: The PSL pebble bed heater shown in this diagram was constructed by a blast furnace manufacturer, Norton Company. The core was lined with harder bricks, and softer bricks were used around the exterior. (NASA)[19]

Image 95: Work on the pebble bed heater's inlet nozzle on 15 November 1958. The heater was active in May, and the tunnel was operational in July. (NASA C–1958–49138)

John Kobak

Kobak worked at Ford and Chevrolet during the summer breaks while pursuing his degree at the Case Institute of Technology. He had intended to accept an automotive plant engineering position after graduating in 1958 but was persuaded by one of his instructors to apply at Lewis. The NACA Lewis lab had a strong affiliation with Case in the 1950s, and his instructor had worked at the NACA.[20]

Like many recent graduates in the late 1950s and early 1960s, John Kobak took a position at the lab on a temporary basis only to become enthralled with the work and spend his entire career there. "I was already prepared to go to Detroit and work for Ford. At the last minute, I decided to come here and stay a few years until I got married and moved on. I liked it so much that I stayed—I stayed my whole career. It was a great place to work. I really enjoyed it."[21] He would spend the next 36 years at the lab.

Kobak used his facilities engineering training by becoming a project engineer stationed at the PSL. "PSL was a really fun place to work.... The research people wrote the reports. I was not a report-type person. I was a hands-on person, so it worked out good."[22]

Initially Kobak worked on rocket installations under Gene Baughman. This included the set of RL-10 engine investigations. By October 1963 Baughman had relocated to Pasadena to oversee Aerojet's work on the M-1 engine, and Kobak assumed his responsibilities at the PSL.

The Centaur Program was transferred to the Lewis Research Center in October 1962. The entire center focused on exorcising the rocket's operating problems. Many of those associated with the RL-10 testing transferred to the high-profile Centaur Program Office, including Kobak's fellow operations engineer Bill Goette. "I chose to remain in the PSL and do that. So like I said, [I made] a lot of bad choices in my career as far as getting promoted sooner. [Goette] became an instant branch chief because he's the only one in the office who knew anything about the engine."[23]

The Airbreathing Engines Division took over the PSL in 1967, and Kobak moved on to the Rocket Laboratory—a series of small test cells that had been used for Lewis's earliest rocket work. He recalled, "[At the Rocket Lab we] worked at all sorts of small engines that nobody cared about. I mean we worked at fluorine mixed with oxygen. We did methane. But still we kept doing ablative stuff. We were doing ablative with hydrogen-oxygen in PSL, and then the program basically continued."[24]

Image 96: John Kobak. (C–1962–64266)

The center entered into new areas of research in the 1970s, however. Kobak and the operations engineers worked for two different divisions on a variety of earth resource programs for the Department of Energy. Former rocket researchers and engineers now focused on projects like the fluidized coal bed at the Rocket Lab. Kobak explained, "Actually I wrote my first report on how to design a whole control system [for the coal bed]."[25]

In 1975 Kobak was named head of the Systems Operations Section of the Chemical Energy Division. His group took over work in the Engine Research Building on the Stirling engine for Chrysler and in setting up a bank of solar arrays for battery testing. Several years of work were completed, but the industry was not interested in the technology. "It was a lot of work but nothing came out of it," Kobak lamented.[26]

In 1987 Kobak undertook his largest and final project at NASA. He was responsible for building the Power Systems Facility, the first facility built specifically for the Space Station Freedom program. The 26,000-square-foot PSF was designed to develop, test, and monitor the space station's electric power system. It contained a 100-foot-square, 63-foot-high class-100,000 clean room. Under Kobak's guidance, the facility was successfully brought online in 1993.

Neal Wingenfeld

Wingenfeld had been a classmate of John Kobak's at the Case Institute of Technology. "He was a fraternity brother at Case," remembered Kobak. "He graduated with me, but he went to Republic Steel and was working there."[27] At a party during the summer of 1963, Wingenfeld let his friends know that he was looking to find a new job.

The Chemical Rocket Division had only one electrical engineer at the time, so Kobak informed his managers that Wingenfeld might be available. "I got a phone call the next week," he reminisced, "and they asked me if I'd come in to interview. And that was the beginning of a process. I ended up with three jobs—two new job offers and my old job." Wingenfeld decided to join his friend at NASA.[28]

Wingenfeld was impressed with the NASA staff. He described them as, "Dedicated, hard-working people, conscientious, very very cooperative. Especially with me. I was the greenhorn on the job so-to-speak there for awhile. It was privilege and a pleasure to work with those people."[29]

"[Fred Looft, the senior electrical engineer at the PSL] was having heart problems," Wingenfeld continued. "They hired me to relieve him off the RL-10 engine project. They didn't want him working at night. Sometimes we would test all night long and have to come back at eight in the morning.... We would test probably two or three times a week. So I took over for Fred, and Fred moved over to Cell 2."[30]

Looft trained Wingenfeld for about six months and continued to be his mentor for several years. "Throughout the Sixties there it was just Fred Looft and myself. We were the only two. So we satisfied all the projects in the division....But Fred and I worked together. We got along very well together. We were office mates."[31]

The RL-10 program was under way when Wingenfeld arrived. "They were having all kinds of problems [with] the controllers [for] the valves that throttled the fuel flows into the engine. And we were doing the gimballing, setting up the control system to do the gimballing and the gimbal controllers. So I immediately got involved with those problems."[32]

Image 97: Neal Wingenfeld. (NASA C–1982–01508)

The RL-10 testing was critical to the nation's space program, and the PSL staff dedicated long hours to resolving its problems. During one stretch Wingenfeld accumulated over 600 hours of compensated time off that he could not use. "You had to be here. And I never went on vacation for three years. I just couldn't get away, was so tied up here I just couldn't get away. It was a hot project. So my first three years here I never took annual leave, and I built up my max instantaneously."[33]

The use of liquid hydrogen was dangerous, but Wingenfeld had spent several years in the perilous conditions at the steel mill, so he was not overly concerned about the PSL. "It's funny, I think the safety people would never allow us to do and operate the way we did back then now. Especially in an unpressurized control room. I never thought much about it. We were so absorbed in what we were doing."[34]

Like Kobak, Wingenfeld left the PSL with the Chemical Rocket Division in 1967. He spent the majority of his next 35 years at NASA Lewis's Rocket Engine Test Facility.

Homemade Rockets

The initial rocket testing at the PSL between 1957 and 1963 involved a number of small rockets and rocket components built by Lewis's fabrication and machine shops. These included rockets fueled by hydrogen peroxide and by isentropic or storable propellants.[35] The first studies were on a water-cooled engine fueled by JP-4 and oxygen. John Kobak explained, "In order to make a large-area-ratio nozzle flow full, you have to [run] it at altitude, or you have to have an ejector that will eject it to that altitude."[36] The PSL afforded this ability.

Researchers and technicians discovered as much from failures as from a successful run. Kobak explained, "The way you learn anything is that something fails, and you figure out how not to have it fail again."[37] The work was both exciting and dangerous. "Fun years for me, for a guy who used to like to blow up things," added Kobak. "We had lots of explosions, lots of blowups."[38]

One 1960 liquid-hydrogen test in PSL No. 1 was particularly memorable. Researchers were testing a Lewis-designed engine in which the cryogenic hydrogen propellant also cooled the nozzle. The test engineers were not sure if the engine needed to be prechilled before it was ignited. Kobak explained, "I was afraid that if we just lit the thing, and we had gas going through there, it would just burn up."[39] Unaware that the engine was built of shock-sensitive material, they decided to precool it.

"Soon as we lit it, it just broke off right there, and it went shooting down the tunnel." Kobak recalled watching the 1960s-era monitors, which had a bit of persistence, "We lost chamber pressure, then the picture is still there with the engine, and then all of a sudden it just faded away."[40]

Image 98: Display showing some of the rocket components manufactured by the Lewis Fabrication Division. (NASA C–1954–35975)

Image 99: John Kobak examines the remains of a failed liquid-hydrogen-cooled engine on 14 June 1960. (NASA C–1960–53702)

Image 100: Monitors in the control room allowed the test operators to view both the liquid-hydrogen storage units and the engine inside the test chamber. (NASA C–1964–69528)

Pursuit of Power

Image 101: Tanking of the Atlas/Centaur-3 rocket at Cape Kennedy on 25 March 1964. The Centaur second stage is behind the white insulation panels near the top. The payload is in the conical nose fairing. (NASA C–1964–68973)

NASA's Propulsion Systems Laboratory No. 1 and 2

Image 102. Ali Mansour and Ned Hannum, engineers in the Chemical Rocket Division, pose with a Pratt & Whitney RL-10 engine on 17 April 1963. (NASA C–1963-64329)

General Dynamics had already begun conceptualizing an upper stage for the Atlas missile. The design was modified in 1957 as a space vehicle and submitted initially to the NACA and then to the U.S. Army. The army accepted the proposal in 1958, and the Centaur rocket was born. The program was supervised by the Army Ballistic Missile Agency in Huntsville (incorporated into NASA in March 1960). In mid-1958 General Dynamics learned of Pratt & Whitney's hydrogen engine work. The RL-10 engines significantly improved the overall Centaur design and performance.[45]

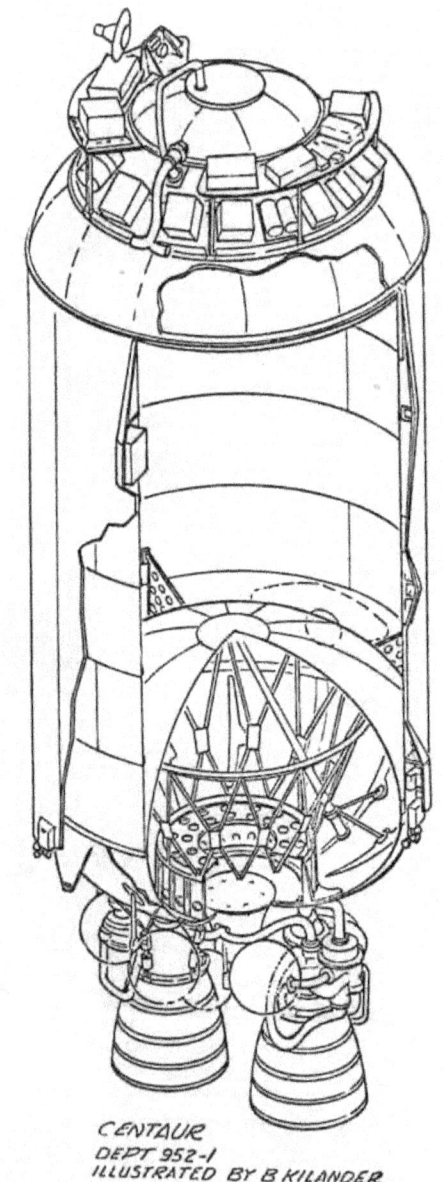

Image 103: Centaur second-stage rocket. (NASA C–1953-34978)

Star of the Rocket Engine Era

NACA Lewis researchers were forced to alter their studies to address the immediate propulsion needs of NASA's space program.[41] For the PSL, this meant investigating performance problems with the Pratt & Whitney RL-10 engine. In 1955 the military had asked Pratt & Whitney to develop hydrogen engines specifically for aircraft. The program was canceled in 1958, but Pratt & Whitney decided to use the experience to develop a liquid-hydrogen rocket engine, the RL-10.[42]

Two of the 15,000-pound-thrust RL-10 engines were used to power the new Centaur second-stage rocket. Centaur was designed to carry the Surveyor spacecraft on its mission to soft-land on the Moon. The criticality of the Surveyor missions to the ensuing Apollo Program resulted in a high profile and tight schedule for Centaur. (For a history of the Centaur Program, see Dawson and Bowles.[43]) "The RL-10 was the star of the rocket engine era here," recalled Neal Wingenfeld. "That was the star program that we had here in PSL."[44]

The Rocket Era   85

Image 104: Lead Test Engineer John Kobak (to the right) and a technician use an oscilloscope to test the installation of an RL-10 engine in the PSL No. 2 on 13 April 1962. (NASA C–1962–60071)

Image 105: Left to right: John Kobak and Gene Baughman examine the first installation of the RL-10 engine in PSL No. 1 on 3 February 1961. The two were in the Rocket and Aerodynamics Division's Rocket Installations Section. Baughman was the operations project engineer for the program. (NASA C–1961–55596)

The RL-10 Gets Worked Over

Centaur was the first cryogenic stage ever attempted and was designed to restart its engines in space. These two features presented a number of design problems. Pratt & Whitney fired the RL-10 over 200 times on a test stand at its Florida facility. On the basis of this success, a series of ignition tests were canceled. The cancellation resulted in two engine explosions the following year.

NASA Headquarters assigned Lewis the responsibility for investigating the RL-10 problems because of the center's long history of liquid-hydrogen development. Lewis began a series of tests to study the RL-10 in March 1960. Two RL-10s exploded during Pratt & Whitney tests in the fall of 1960 and spring of 1961.[46]

The first series of RL-10 tests in early 1961 involved gimballing, or steering, the engine as it fired. The engine mount system had a bearing system that allowed the engine to rotate up to 2° in either direction. Pratt & Whitney was able test the engine at their facility in Florida, but their setup required the rocket to be direct-connected to the exhaust ejector. They were unable to gimbal the engine.[47] At Lewis, researchers were able to yaw and pitch the engine in the PSL as it was firing to simulate its behavior during a real flight. Over the course of several months, they were able to test and evaluate the performance of the gimballing system.

Pratt & Whitney altered the original RL-10 design to allow throttling to produce different levels of thrust. Again, the PSL was needed to test the engine. A team of Chemical Rocket Division researchers studied the RL-10's performance when throttled from 1,500 to 15,000 pounds of thrust. Despite some low-frequency fuel oscillations, they found that, overall, the engine performed well throughout the throttling range.[48]

Kobak remembered, "We burned out some of the engines finding problems. We found problems with the engines at low thrust going into this chugging mode, and we had to develop a way to prevent that."[49]

This led to another key improvement of the RL-10 in the PSL, the resolution of the low-frequency combustion instability in the fuel system, or chugging. In most rocket engine combustion chambers, the pressure, temperature, and flows are in constant flux. The engine is considered to be operating normally if the fluctuations remain random and within certain limits.[50] Lewis researchers used high-speed photography to study and define the RL-10's combustion instability by throttling the engine under the simulated flight conditions. They found that the injection of a small stream of helium gas into the liquid-oxygen tank immediately stabilized the system.[51]

The low-frequency oscillations at low thrust levels were a little more difficult to resolve. Ultimately it was determined that the abrupt change in the propellant's density as its temperature increased in the cooling jacket caused the instability. This was combated with the injection of gaseous helium or hydrogen just upstream from the cooling jacket.[52]

Another issue with chemical rocket engines is screech, an increasingly loud tone that causes a damaging rise of component temperatures. Screech is caused by the speed of gas oscillations during combustion. These oscillations erode the boundary layer until the temperature of the walls begins increasing. Screech can severely melt or burn out the rocket components in a very short amount of time. Lewis researchers explored various methods of acoustic damping to suppress the oscillations. The use of resonators to successfully reduce screech was demonstrated in the PSL.[53]

In order to keep the engine system at cryogenic temperatures before launch, the engine's turbopumps were flooded with liquid hydrogen. Kobak explained, "So they're wasting a full-flow of liquid hydrogen, which is not going to help the mission. They're just wasting it before they light the engine."[54] Lewis researchers decided to try to cool the pump with helium and wait to flow the hydrogen into the engine until it was time to ignite.[55] Insulation was also installed to keep the system cold until the upper stage was ignited. The method of chilling the system with helium before launch was first demonstrated in the PSL on an RL-10. This precooling was one of Lewis's most important modifications to Centaur and is still used today.[56]

The NASA Marshall Space Flight Center initially managed the Centaur and RL-10 programs but generally did not care for the spacecraft's balloonlike structure or the use of liquid hydrogen as the propellant. Marshall Director Wernher von Braun called for the cancellation of the program even before the failure of the first launch in May 1962. Congressional hearings and internal debate at NASA lasted most of the summer. In September the Centaur Program was transferred to Lewis. Marshall retained oversight on the RL-10 program until 1966.

One of the primary reasons for the transfer was that the PSL had been testing the RL-10 engines since 1961. Former Centaur manager Larry Ross explained, "[Center Director Abe Silverstein] knew that we were running RL-10s out of PSL, hot-firing RL-10s out of PSL. So he had a great deal of confidence that we had the kind of expertise that it took to make Centaur work reliably. And he's right; there were the kind of people here who could do it."[57]

Sitting on Top of All That Hydrogen

PSL's liquid hydrogen and liquid oxygen were stored in temperature-controlled tanks outside the Shop and Access Building. Nitrogen, which was used to move the propellants through the lines, was trucked in to the site.

Early in the RL-10 test program, liquid hydrogen was still used to prechill the engine. This initial blast of hydrogen would not always be completely burned when the engine was ignited. One time some of the fuel passed through the coolers and exhaust system and into the Equipment Building. The resulting explosion damaged the massive equipment. Operations engineers quickly designed a system that vented that initial flow of hydrogen to the PSL roof and burned it off into the atmosphere.[58]

Image 106: Liquid-oxygen and liquid-hydrogen tanks brought in for a PSL rocket test on 16 April 1963. The control room is inside the building to the right of the awning. (NASA C–1963–64268)

Image 107: An engineer monitors the high-pressure fuel-pumping system in the rear of the PSL control room on 9 December 1963. (NASA C–1963–67414)

Image 108: Several large-area-ratio nozzles are lined up on 1 May 1964 in front of PSL No. 1 for tests with "storable" propellants. Nozzles (left to right): 15° conical, Apollo Space Propulsion System contour, and U.S. Air Force Titan transtage contour. The veteran mechanic Bill Neidengard is examining the nozzles with a young researcher. (NASA C–1964–69633)

John Kobak recalled, "[It was] a lot of fun and we were so absorbed trying to do a good job that we didn't think of the dangers. Until later on when people were saying, 'You were sitting on top of all that hydrogen and oxygen.' Those tanks were right outside, the control room's right there. I mean now, like up at Plum Brook, the control room for B-2 is like half a mile away. We were fifty feet away."[59]

In May 1966 the RL-10 program was officially transferred to Lewis from Marshall and placed under the Centaur Program Office. By that time, Lewis had successfully resolved most of Centaur's problems and turned it into a reliable space vehicle. After six Lewis-managed Centaur test missions, the first Surveyor launch attempt on 2 June 1966 was successful. Centaur performed perfectly, and Surveyor became the first spacecraft to land on another celestial body.

Rocket Division Arrives

The Chemical Rocket Division was created in 1963 and took over all testing in the PSL. Its immediate predecessor at the PSL, the Propulsion Systems Division, contained only 25 engineers and 25 technicians in 1956.[60] Division Chief Irving Johnsen managed over 100 engineers scattered across 15 sections. Frequent PSL researchers Ned Hannum, Ali Mansour, and John Wanhainen were in the Engine Systems Section under Harry Bloomer. Others such as Carl Auckerman and Art Trout were in Carl Meyer's Engine Components Section. Neal Wingenfeld, Kobak, and PSL No. 2 test engineer Wayne Thomas were in the Engine Operations Section under Al Ross.

The Test Installations Division was still responsible for installing the equipment and running the facility. Bud Meilander continued to oversee the PSL mechanics.

Primitive Propulsion

By the early 1960s the development cost of advanced chemical rockets was increasing to the point of being prohibitive. It was not uncommon to retest large engines 50 to 100 times to investigate modifications. Lewis's role was to study broader concepts such as materials, cooling, propellant feed systems, and combustion instability. NASA sought to improve rocket engine performance while reducing size, costs, and development time for the entire industry.[61]

In this vein, Lewis conducted a wide-ranging storable propellant program during late 1963 and 1964 that focused on a 9,000-pound-thrust engine. Storable propellants, such as nitrogen tetroxide and hydrazine, are appealing because they do not require any special temperature- or pressure-control measures. NASA planned on using storable propellants for the upper stages of the Saturn V rocket. Although this type of fuel had been studied for several years, combustion instability, ablative thrust chamber durability, and nozzle efficiency had to be investigated before storable propellants could be used for the space program. The PSL was selected to conduct these investigations because of its ability to test large engines in an altitude environment.[62]

Researchers sought to determine the impulse value of the propellant mix, improve the performance, and correlate the results with other analytical tools. A number of ejector and combustion chamber designs were studied with both low- and high-area-ratio nozzles. Two contour nozzles were machined specifically for the PSL.[63] In addition, two specially manufactured contour nozzles—one from the Apollo Space Propulsion System and one for the U.S. Air Force Space Systems Division's Titan transtage engines—were tested with various injectors.[64]

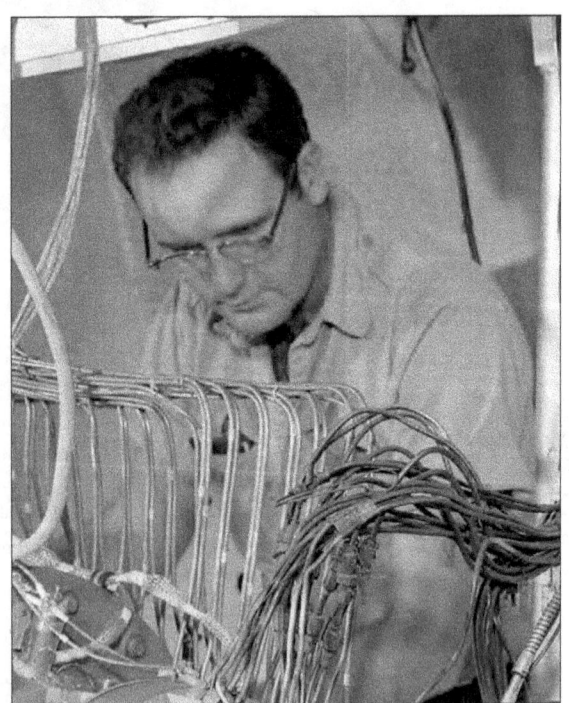

Image 109: Wayne Thomas in PSL No. 2. (NASA C-1958-49066)

By not using a flame collector tube, NASA researchers were able to forgo the usual problems of firing low-area-ratio nozzles in an altitude chamber. For the high-area-ratio nozzle tests, a 72-inch-diameter diffuser tube was used to exhaust the flame.

The researchers found that the nozzle performance was unaffected by combustion instability, but that the measurement of combustion chamber pressures was not a reliable indicator of the chamber or nozzle performance. Although valuable information was obtained during the tests, attempts to improve the engine performance were not successful.[65]

NASA became increasingly interested in solid rockets for heavier payloads as the costs of the space program escalated in the mid-1960s. Solid rockets are significantly less complicated and expensive than chemical rockets. They contain only a nozzle, the propellant, and an igniter. The rocket's shell serves as the pressure chamber.[66] Not all payloads required the sophisticated rocket concepts originally developed for missiles. Solid rocket proponents believed that larger, "dumber" rockets could perform many of the same missions at a much lower cost.[67]

In June 1963 Aerojet began developing a 260-inch-diameter engine for the air force at its specially constructed test facility in the Florida Everglades. In March 1965 NASA assigned Lewis the responsibility for a feasibility study of Aerojet's 3.25-million-pound-thrust rocket. The massive rocket had 260-inch-diameter nozzles and contained 1.6 million pounds of solid propellant.[68]

The lack of existing data caused Aerojet to struggle with the steering system for the motor's aft end. Aerojet systematically created these data as well as an analytical model to predict the ignition motor pressure levels.[69] Shortly thereafter, Lewis researchers Salmi Reino and James Pelouch performed a series of small-scale model tests of the engine in the PSL.

Reino and Pelouch decided the best method to improve the aft thrust vector control was to have a gimballing nozzle. The new design relied on a bearing that sat directly in the nozzle's smallest diameter area and the insertion of the nozzle further into the actual engine. The new hot-gas flow patterns raised concern regarding the nozzle's structural integrity, particularly in the high-velocity annular region.

During the summer of 1966 Lewis used a 0.07-scale model of the engine in the PSL at altitude conditions with compressed airflow to study the velocity and direction of the annular channel flow. High-speed color motion pictures helped researchers to determine the causes of the flow. After several attempts, Lewis was able to modify the model so that the nozzle's integrity was bolstered by increased insulation and the annular airflow was lower than in the design before gimballing. This configuration was verified during ambient conditions inside the former Altitude Wind Tunnel.[70,71]

Image 110: An Apollo contour engine is installed in PSL No. 1 on 13 March 1964. Cylindrical and conical combustion chambers were used on the engines. Large-area-ratio nozzles with 15° conical skirts could be used with any of the combustion chambers. Two contour nozzles were machined specifically for a particular chamber. Several injectors were studied, and three were selected for in-depth investigation. (NASA C–1964–68789)

NASA's Propulsion Systems Laboratory No. 1 and 2

Image 111: A full-size Aerojet 260-inch-diameter rocket and its motor case during a 25 October 1965 investigation of a hydrostat failure. Shipbuilders were contracted to construct the unique large engine chambers.⁷² (NASA C–1965–03064)

Image 112: Bill Neidengard inspects the PSL No. 2 installation of a scale model of a large high-energy solid rocket engine in 1963. The majority of the nozzle was cut away to facilitate instrumentation.⁷³ (NASA C–1965–1037)

The Rocket Era

Pursuit of Power

By 1966 two half-scale versions of the motors had been fired at an Aerojet facility in Florida, generating 1.6 million pounds of thrust. The 2.67 million pounds of thrust generated by a third engine test was the largest amount of thrust ever produced by any type of rocket.[74] A full-scale, nongimballing test was scheduled for mid-1967.[75]

The program successfully ended in 1967, and several new cost-saving technologies were demonstrated. Lewis continued with other solid rocket activities to further reduce launch costs, investigate problem areas, and improve reliability. Like other space technologies that had no immediate application, however, the 260-inch-diameter rocket program was canceled as NASA's budgets dried up in the late 1960s.[76]

End of an Era

The NACA Lewis 1957 Inspection signaled the start of Lewis's space program, and the 1966 Inspection marked the end. The talks at the PSL station dealt almost exclusively with the advanced chemical rocket studies of the previous eight years. The RL-10 studies produced critical advances that led to the success of the Centaur rocket. A General Electric J85 jet engine installed in PSL No. 2, however, portended the impending transition of both the PSL and the NASA Lewis Research Center back to airbreathing engine work.

Image 113: Setup for a PSL Inspection talk on 7 October 1966. A Pratt & Whitney RL-10 engine is at the left of the stage, an injector plate is at the right, and a GE J85 turbojet is in the test cell at the far right. (NASA C–1966–03902)

Endnotes for Chapter 5

1. Neal Wingenfeld interview, Cleveland, OH, 15 September 2008, NASA Glenn History Collection, Oral History Collection, Cleveland, OH.
2. John Kobak interview, Cleveland, OH, 1 September 2009, NASA Glenn History Collection, Oral History Collection, Cleveland, OH.
3. Virginia Dawson, *Engines and Innovation: Lewis Laboratory and American Propulsion Technology* (Washington, DC: NASA SP–4306, 1991).
4. Virginia P. Dawson, "Ideas Into Hardware: A History of the Rocket Engine Test Facility at the NASA Glenn Research Center," NASA Glenn, Cleveland, OH, 2004, http://history.nasa.gov/retfpub.pdf.
5. John Sloop, *Liquid Hydrogen as Propulsion Fuel, 1945–59* (Washington, DC: NASA SP–4404, 1978).
6. V.F. Hlavin, E.R. Jonash, and A.L. Smith, *Low Pressure Performance of a Tubular Combustor With Gaseous Hydrogen* (Washington, DC: NACA RM–E54L30A, 1955), pp. 9–10.
7. Dawson, *Engines and Innovation*, chap. 8.
8. Dawson, *Engines and Innovation*, chap. 8.
9. "1957 Inspection October 7–10, 1957," *Wing Tips* (11 September 1957).
10. Howard Wine interview, Cleveland, OH, 4 September 2005. NASA Glenn History Collection, Oral History Collection, Cleveland, OH.
11. Propulsion Systems Division, "Data Tabulation, Altitude Chambers, PSL-1."
12. Propulsion Systems Division, "Data Tabulation, Altitude Chambers, PSL-1," NASA Glenn History Collection, Test Facilities Collection, Cleveland, OH.
13. M. J. Krasnican, et al., "Propulsion Research for Hypersonic and Space Flight," NACA Lewis Flight Propulsion Laboratory Inspection, 7–10 October 1957, NASA Glenn History Collection, Cleveland, OH.
14. Eugene Manganiello to I. H. Abbott and E. W. Conlon, 20 January 1960, NASA Glenn History Collection, Cleveland, OH.
15. Krasnican, et al., "Propulsion Research for Hypersonic and Space Flight," p. 4.
16. Hypersonic Tunnel Facilities," 23 September 1957, NASA Glenn History Collection, Director's Collection, Cleveland, OH.
17. Robert Walker interview, Cleveland, OH, 2 August 2005, NASA Glenn History Collection, Oral History Collection, Cleveland, OH.
18. Propulsion Systems Division, "Data Tabulation, Altitude Chambers, PSL-1."
19. Nancy Barron, "Pebble Bed Enables High-Temperature Research," *Wing Tips* (26 February 1958).
20. Kobak interview, 2009.
21. Kobak interview, 2009.
22. Kobak interview, 2009.
23. Kobak interview, 2009.
24. Kobak interview, 2009.
25. Kobak interview, 2009.
26. Kobak interview, 2009.
27. Kobak interview, 2009.
28. Wingenfeld interview, 2008.
29. Wingenfeld interview, 2008.
30. Wingenfeld interview, 2008.
31. Wingenfeld interview, 2008.
32. Wingenfeld interview, 2008.
33. Wingenfeld interview, 2008.
34. Wingenfeld interview, 2008.
35. Propulsion Systems Division, "Data Tabulation, Altitude Chambers, PSL-1."
36. Kobak interview, 2009.
37. Kobak interview, 2009.
38. Kobak interview, 2009.
39. Kobak interview, 2009.
40. Kobak interview, 2009.
41. Virginia P. Dawson, "Ideas Into Hardware."
42. Dawson, *Engines and Innovation*.
43. Virginia Dawson and Mark Bowles, *Taming Liquid Hydrogen: The Centaur Upper Stage Rocket, 1958–2002* (Washington, DC: NASA SP–4230, 2004).
44. Wingenfeld interview, 2008.
45. Dawson and Bowles, *Taming Liquid Hydrogen*.
46. R. F. Holsten, "Centaur Propulsion Systems Testing Volume I," 7 December 1962 (Pratt & Whitney FR–431).
47. Kobak interview, 2009.
48. John P. Wanhainen, et al., *Throttling Characteristics of a Hydrogen-Oxygen, Regeneratively-Cooled, Pump-Fed Rocket Engine* (Washington, DC: NASA TM X–1043, 1964).
49. Kobak interview, 2009.
50. "Advanced Chemical Rockets: Presented at 1966 Inspection of the Lewis Research Center," 4 October 1966, NASA Glenn History Collection, Inspections Collection, Cleveland, OH.
51. William E. Conrad, Ned P. Hannum, and Harry E. Bloomer, *Photographic Study of Liquid-Oxygen Boiling and Gas Injection in the Injector of a Chugging Rocket Engine* (Washington, DC: NASA TM X–948, 1964).
52. Conrad, Hannum, and Bloomer, *Photographic Study of Liquid-Oxygen Boiling and Gas Injection*.
53. "Advanced Chemical Rockets."
54. Kobak interview, 2009.

55. Kobak interview, 2009.
56. "1962–1972: The First Decade of Centaur," *Lewis News* (20 October 1972).
57. Larry Ross interview, Cleveland, OH, 1 March 2007, NASA Glenn History Collection, Oral History Collection, Cleveland, OH.
58. Kobak interview, 2009.
59. Wingenfeld interview, 2008.
60. Propulsion Systems Division, "Data Tabulation, Altitude Chambers, PSL-1."
61. "Advanced Chemical Rockets."
62. Carl Auckerman and Arthur Trout, *Experimental Rocket Performance of Apollo Storable Propellants in Engines With Large Area Ratio Nozzles* (Washington, DC: NASA TN D–3566, 1966).
63. Auckerman and Trout, *Experimental Rocket Performance of Apollo Storable Propellants*.
64. Auckerman and Trout, *Experimental Rocket Performance of Apollo Storable Propellants*.
65. Auckerman and Trout, *Experimental Rocket Performance of Apollo Storable Propellants*.
66. "Advanced Chemical Rockets."
67. Lionel Johns, Ray Williamson, and Richard DalBello, "Big Dumb Boosters: A Low-Cost Space Transportation Option?: AN OTA Background Paper," February 1989, Workshop on Low-Technology, Low-Cost Space Transportation Options, 1 December 1987.
68. *260-Inch Diameter Motor Feasibility Demonstration Program* (Sacramento, CA: Aerojet General Corporation, 1966).
69. Carl Ciepluch, *Status of the 260-Inch Diameter Solid Rocket Motor Program* (Washington, DC: NASA TM X–52275, 1967).
70. Reino Salmi and James Pelouch, Jr., *Investigation of a Submerged Nozzle for Solid Rockets* (Washington, DC: NASA TM X–52285, 1967).
71. Reino J. Salmi and James J. Pelouch, Jr., *1/14.2-Scale Investigation of Submerged Nozzle for SL-3 260-Inch Solid Rocket* (Washington, DC: NASA TM X–1546, 1968).
72. "Advanced Chemical Rockets."
73. Salmi and Pelouch, *Investigation of a Submerged Nozzle for Solid Rockets*.
74. "Advanced Chemical Rockets."
75. Salmi and Pelouch, *Investigation of a Submerged Nozzle for Solid Rockets*.
76. J. D. Hunley, *The Development of Propulsion Technology for U.S. Space-Launch Vehicles, 1926-1991* (College Station, TX: Texas A&M University Press, 2007), p. 270.

Image 114: A close-up of the Pratt & Whitney TF30 turbofan engine with instrumentation installed on 11 August 1967. (NASA C–1967–02827)

· 6 ·
Jet Engines Roar Back

Abe Silverstein made a daring move in January 1966 that would transform the NASA Lewis Research Center for the next decade. Although the Apollo Program was ramping up, and the nation was captivated with the space program, Silverstein decided to steer the center away from space research and back into aircraft propulsion. The sharp break had its antecedents at the center with equally harsh adjustments for the jet engine in 1945 and the space program in 1957. Silverstein felt that individuals performed best when thrown into unfamiliar areas.

"So many people, when they get immersed in a given field, aren't willing to move out of it because it becomes comfortable," Silverstein later explained. "And if we hadn't done it sharply, we would have been slipping, sliding around in a lot of old projects for a long time and losing effectiveness on things that weren't really new and going anywhere."[1]

The move made some anxious because the staff had been consumed by space activities for the last decade, and the Apollo Program was about to come to fruition. In addition, the center's powerful test facilities, including the Propulsion Systems Laboratory (PSL), now found themselves competing with those in private industry and at the Arnold Engineering Development Center (AEDC).[2,f,3]

A host of new aeronautical challenges had arisen during the previous decade. Europeans were making strong advances with the long-awaited supersonic transport (SST) technology. The French prototype was being assembled at the time and the Russian SST would make its initial flight in 1968.[4] Meanwhile, the United States struggled vainly throughout the 1960s to develop its own version.

Unlike their groundbreaking engine work in the 1940s and 1950s, Lewis's new studies were not only for the military but also for the Federal Aviation Administration and the Department of Transportation. The increase in commercial flights and airport congestion resulted in new problems that needed Lewis's expertise, including noise abatement and emissions reduction.

The revolutionary growth of the airlines in the mid-1950s was powered by a new type of engine: the turbofan. Turbofans pass most of the airflow around the engine core. This reduces noise levels, decreases fuel consumption, and generates additional thrust at low velocities. The turbofan was more efficient and powerful than the turbojet, but it had its own set of design problems. In addition, modern engines were increasingly susceptible to compressor flutter and required more sophisticated control systems. Clearly, there was an abundance of new problems for Lewis to resolve.

Silverstein outlined some of the issues facing engine manufacturers during a September 1966 talk in London. He credited the expansion of the airline industry to the lighter, longer-lasting, and more efficient engines, then went on to describe some of the problems.

"These requirements lead to complex engine machinery which must be compromised to provide, not only effective and economical performance and operations, but which must also operate quietly. Further, the need for efficiency generally conflicts with the requirement that the aircraft be tolerant of [the] flight environment and forgiving of error in design, construction, and handling."[5]

---

[f] Herbitus, Germany's first modern altitude chamber and the first altitude testbed for jet engines, was operating by 1944. At the end of the war, the Herbitus was dismantled by American troops and shipped to the United States. It was reassembled at the new AEDC in 1954.[3]

Airbreathing Research Personnel

After 10 years of rocket studies, the PSL would play a major role in the new aeronautical program. In 1967 Lewis created the Airbreathing Engines Division to address these new propulsion issues. Its Chief, J. Howard Childs, was among the first to work on jet propulsion at the Cleveland lab in 1943. The Airbreathing Engines staff, which consisted entirely of operation engineers, was housed in the PSL Operations Building. Many of the PSL Chemical Rocket Division personnel were integrated into the new division and remained at the facility.

Others, like John Kobak and Neal Wingenfeld, remained with the rocket group and moved on to other facilities.[6] Howard Wine remembered being reassigned to the new Zero Gravity Facility. "[I] knew every nut and bolt in the place [PSL] for I think about 10 years and all of a sudden my boss came to me and said, 'we would like you to go over and kind of get in on the ground work of Zero-G'.... They asked me if I would be interested in going over there. I said, 'yeah,' even though I hated to leave PSL."[7]

Frank Kutina oversaw the section of the Airbreathing Engine Division's Research Operations Branch that worked with the PSL researchers to develop the test configurations and setup.[8] The Test Installations Division's mechanics, electricians, and electrical technicians continued to be responsible for building up the test facilities and setting up tests. The Test Installations Division personnel worked closely with Kutina and his crew.

The General Electric (GE) J85 airflow distortion program elucidated the interaction of the different divisions for a single PSL investigation. The Airbreathing Engines Division was responsible for designing and fabricating the instrumentation, preparing the

Image 115: A Pratt & Whitney TF30 turbofan engine about to be installed in PSL No. 1 on 6 October 1967. Left to right: PSL Electrical Engineer Fred Looft and Engineering Services Chief Robert Godman. (NASA C–1967–03560)

Image 116: Diverse types of engineers from several divisions coordinate to install a GE engine in PSL No. 1 on 8 May 1967. (NASA C–1967–01584)

Image 117: Frank Kutina, seen here in 1957, served as head of PSL operations engineers a decade later. He retired in 1994 with 53 years of NASA service and an Exceptional Service Medal. (NASA C–1957–44592)

engine, overseeing its installation into the test cell, and providing support during the test runs. The Fluid System Components Division selected and fabricated the experimental equipment and consulted on the instrumentation and data acquisition.

The Wind Tunnels and Flight Division coordinated the PSL testing with other facilities, selected the experimental equipment, and provided the project engineer.[9] The Test Installations Division technicians worked with these groups to integrate the equipment for the test into the facility.

An emerging trend during this period was the study of different research topics using a single engine already under investigation for different purposes. Over the next decade, the GE J85 and Pratt & Whitney's F100 and TF30 would each be used for a variety of different tests.

Pursuit of Power

Image 118: Bud Meilander views the flamespreader in PSL No. 2 on 18 May 1967. As the engines being tested in the PSL increased in size, the primary cooler became damaged by the higher temperature exhaust flows. The flamespreader slowed the flow of hot gases at the cooler inlet by spreading the exhaust over a larger area. This increased the cooler's heat-transfer capabilities and prevented damage to the cooler tubing.[10] (NASA C–1967–01733)

NASA's Propulsion Systems Laboratory No. 1 and 2

Image 119: Lewis's new F-106B Delta Dart aircraft on 14 June 1968 showing the two research nacelles with J85 engines beneath its delta wings. One nacelle housed a "reference engine" and the other the modified test engine. Both were operated under identical conditions and had identical drag profiles. Load cell differences could then be attributed to characteristics of the research engine.[11] (NASA C–1969–02871)

Supersonic Calibration

Boeing led the nation's effort in the late 1960s to develop an SST aircraft powered by four massive 65,000-pound-thrust GE4 turbojets. The amount of fuel required to reach supersonic speeds and engine noise levels were two of the larger problems besetting the program. In order to provide a better general understanding of these issues, the NASA Lewis Research Center undertook extensive drag and noise-reduction studies from 1967 to 1977. The programs focused on the inlets and nozzles for advanced propulsion systems.

Jet engines can be deafening, but only a small amount of a turbojet's in-flight noise is created by the engine itself. The primary cause is the interaction of the hot exhaust gas with the surrounding atmosphere. Lewis had first studied the use of noise-suppressing nozzles and ejectors in the late 1950s. The new studies focused on nozzle performance in the transonic range. The goal was to determine the change in noise and thrust created by supersonic airflow at the exhaust nozzle.[12]

GE's relatively small but potent J85-13 turbojet was selected for this study because its "reasonable size" enabled it to be mounted under the wings of the flight research vehicle and reduced fuel consumption during the tests.[13] GE still claims that the 18-inch-diameter J85 provides the highest thrust-to-weight ratio of any engine in its class.[14]

Two GE engineers traveled to Cleveland in May 1967 to prepare the engine for its initial runs in the PSL. The men praised the technicians for their setup of the PSL chamber. They claimed it was one of the best tank installations they had witnessed despite Lewis's paucity of recent jet engine work. The GE and Lewis engineers proceeded to collaborate on several small improvements to the installation throughout the summer.[15]

Different nozzle configurations were first explored in the center's two large supersonic wind tunnels. Because the tunnels' size limitations precluded complete engine testing, a full-scale version was then checked out in the PSL. The modified engine was then flown at transonic and supersonic speeds on the F-106B. Mounting the research engines under the F-106B's delta wings provided researchers with a sub-scale facsimile of the SST.[16]

Jet Engines Roar Back   103

The 1968 and 1969 runs in the PSL ensured flight worthiness, verified the performance of the engine and nozzle, and determined pressure and temperature profiles.[17] This engine calibration was performed prior to any modifications to the engine's noise-reduction treatment, nozzle, or compressor. The engine's inlet pressure, temperatures, and rate of gas flow had to be determined at transonic speeds in order to define the characteristics of a nozzle.[18]

William Latto, Jr., a veteran of Lewis's early rocket work, was able to use the PSL test data to calculate the engine's inlet thrust and specific fuel consumption within a 0.06-percent error margin.[19,20] These measurements served as the basis for future engine and nozzle tests in the PSL, on the center's F-106 research aircraft, and in the supersonic tunnels.[21]

Over the course of forty F-106B research flights in 1970, six variable-geometry supersonic-cruise variations and two inlets were analyzed.[22,23] These included variable-flap ejectors and plug and auxiliary inlet ejector nozzles. The plug nozzle was first acoustically tested while the aircraft sat on the airport tarmac.

Afterward, a series of passes were made 300-feet above the audio-recording equipment.[24] At subsonic cruise, the variable-flap ejector appeared to be the most promising, with the plug nozzle next best. However, from ground test results, the plug nozzle appeared to be less noisy than others.[25] The regimen continued in 1971 and included a plug nozzle cooled by compressor bleed air.[26]

Boeing's SST never made it past the drawing board. Congress canceled the program in October 1971 as the public became increasingly wary of its noise, sonic booms, and the possible ozone depletion over U.S. territory. The SST was not economical to fly at the subsonic speeds required to reduce noise levels. Interestingly, the Federal Aviation Administration approved the Concorde for overland flights weeks after canceling Boeing's program.[27]

Although Lewis's SST noise-reduction efforts were disappointing overall, researchers did make abatement strides by using the plug nozzle and relocating the engine underneath the wing.[28] The F-106B completed its last flight for Lewis on 8 January 1977. In the eight

Image 120: Lee Wagoner examines the 4,080-pound-thrust J85-13 setup in PSL No. 2 on 26 May 1967. GE engineers had recently arrived to assist with the installation. (NASA C–1967–02128)

and a half years at Lewis, the aircraft flew over 300 test flights, and the research contributed to 45 technical reports.[29]

Artificial Distortion

The design and performance of aircraft engines and inlets continually evolved to meet escalating expectations. The constant battle to increase thrust while decreasing overall weight created additional stress on jet engine components, particularly compressors. As speed and maneuverability were enhanced, the strain on the engines and inlets grew. This strain was primarily a result of inlet flow distortions and lower Reynolds numbers. This thorny combination reduced compressor stability and led to increased stall margins.[30]

The distortions are produced by shifts in either pressure or temperature. Generally, pressure distortions are generated by strong winds, high angles of attack, aircraft wakes, or boundary layer interactions. The temperature distortions, which are less common, arise from the exhaust of other aircraft or the firing of weapons.[31]

Image 121: A Pratt & Whitney TF30 engine being prepared on 10 August 1967 for a test in PSL No. 1. (NASA C–1967–02809)

In 1968 Lewis undertook a wide-ranging, long-term study of airflow distortion. The goal was to collect a large amount of data and combine it in analytical models. The 10- by 10-Foot Supersonic Wind Tunnel was used to study the compatibility and control of different inlets and engines under simulated flight conditions at the compressor. The PSL was more suitable for simulating the conditions at the engine inlet.[32]

Two very different engines would be the primary testbeds for the distortion program—the GE J85-13 turbojet and the Pratt & Whitney TF30 turbofan. The TF30, which emerged in the early 1960s as the nation's first 25,000-pound-thrust engine, dwarfed the J85 in size and power. The TF30 powered the U.S. Air Force's F-111 Aardvark and the U.S. Navy's F-14A Tomcat fighter jets. The 4,080-pound-thrust J85 became one of GE's most enduring and successful engines, particularly on the Northrop T-38 Falcon and F-5 Freedom Fighter.

Lewis conducted a number of tests in 1969 and 1970 to examine the airflow distortion on the Pratt & Whitney TF30-P-1 and TF30-P-3 engines in PSL No. 1. Similar studies were run concurrently on the GE J85-13 in PSL No. 2. The engines were first tested without any distortion to establish baseline parameters.[33]

Engineers in the PSL employed different methods of distorting the airflow to simulate disturbances found during flight. During the early studies, screens were inserted into the inlet duct to create the distortions. The TF30-P-1 compressor proved impervious to four different flow-distortion patterns used in one study.[34]

Altering the screen setup was a time-intensive chore, however. Lewis researchers devised a distortion device in PSL No. 1 with a series of 54 air jets that systematically fired into the airstream. A test of the new system on the TF30-P-1 demonstrated that the air jets were just as effective as the screens, and by their very nature, the air jets were more versatile and easier to install.[35]

The Lewis researchers were then able to map the engine's likeliness to stall when it was subjected to different pulses or distortions. It was found that the duration of a stall-inducing pulse was the inverse of its amplitude.[36,37]

A TF30-P-3 also underwent an overall checkout in PSL No. 1. Attention was focused on the afterburner performance and engine operating limits.[38] The PSL tests established that the increased speeds and altitudes produced inlet flow distortions that, along with decreased Reynolds numbers, reduced the stability of the engine's compressor. In addition, afterburner performance decreased as altitude increased.[39]

During the summer of 1969 a gaseous hydrogen burner was set up in front of both the TF30-P-1 and TF30-P-3 to create temperature disturbances. The individual sections of the burner were independently controlled to produce a number of different distortion patterns. The researchers discovered that inlet temperature distortion had a significant effect on engine stability.[40] Compressor stall occurred with temperature changes of 14 to 20 percent.[41]

The third phase of the studies analyzed different combinations of pressure and temperature distortions. The systematically mapped engine results showed effects similar to those predicted by computer models.[42]

Treatments

The distortion research provided a better understanding of the effect of transient distortions on engine behavior.[43] The PSL studies were performed in parallel with Lewis's computer modeling efforts. The engine tests were used to refine the virtual models, which the flow specialists used to develop strategies for combating inlet distortion. One technique involved carving slots or grooves in the compressor casing to guide the tips of the compressor blades. This was referred to as "treatments."[44]

The Lewis researchers who studied various casing treatments on single-stage compressors found that the treated compressors had increased flow range, distortion tolerance, and operating envelope. Leon Wenzel led a team that investigated the effects of the treatments on multistage compressors at the PSL. It was a cooperative program involving the Airbreathing Engine, Fluid System Components, and Wind Tunnels and Flight divisions.[45]

The Lewis team selected the J85-13 for the investigation because of the staff's familiarity with the engine. GE thinned the engine's stator segments and designed

Image 122: Leon Wenzel (seen here in 1957) of the Systems Dynamics Section in the Wind Tunnels and Flight Division led the compressor casing treatment studies. (NASA C–1957–45544)

an almost entirely new casing to accommodate the treatments.[46] Stress fatigue studies confirmed that the modified blades were as strong as GE's original stator design.[47] The casing included removable inserts so that different treatment types could be installed and tested. GE provided three sets of treatments—an angled blade slot, a circumferential groove, and no treatment.[48]

In late 1973 and 1974 Wenzel's researchers examined the treatment types, the optimal combination of compressor stages to treat, and possible decreases in efficiency in PSL No. 2. Each of the engine's eight compressor stages was individually instrumented to provide undistorted inlet conditions.[49,50]

The researchers began by mapping the engine with and without treatments for the blade tips. Then all eight stages were outfitted with circumferentially grooved rings and were tested. The engine was run again with the rings only on the last three stages. Rings with angled slots were then tested in a similar fashion. The PSL studies, however, found that the treatments did not improve performance and that, in some cases, they actually reduced pumping capability.[51,52] Research into the use of treatments has continued over the years, and they have been incorporated into some engine designs with some degree of success.

Slower, Better, Cheaper

One of the new initiatives emerging from Lewis in the late 1960s was a small 650-pound-thrust low-cost jet engine. Jet engines had proven themselves on military and large transport aircraft, but cost precluded their use on small general aviation aircraft. Lewis undertook a multiyear effort to develop a less expensive engine to fill this niche.[53]

The Low Cost Engine was to be 75-percent less expensive than normal jet engines and was to use less fuel and emit fewer pollutants than did the reciprocating engines then in use. In addition, they would increase the aircraft's speed, range, and safety.[54] Performance sacrifices were necessary because of the temperature and pressure limitations required to fabricate the engine from low-cost materials.[55]

"Our mission was not to invent a new type of engine," explained project manager Harold Gold, "but to find ways to simplify small gas turbine engines, reducing the production costs."[56]

The navy became interested in using the technology as a possible alternative to the rockets that powered their expendable drone aircraft. The navy began cosponsoring the program in 1970, and Lewis altered the engine design to meet their specifications.[57]

The four-stage, axial-flow Low Cost Engine was constructed from sheet metal. It was only 11.5 inches in diameter and weighed 100 pounds.[58] The final design specifications were turned over to a manufacturer in 1972. Four engines were created, and as expected, the fabrication and assembly of the engine were comparatively inexpensive.[59]

During sea-level tests on a test stand in the Special Projects Laboratory, low compressor efficiencies prevented the Low Cost Engine from meeting

Image 123: Researchers Robert Cummings and Harold Gold on 2 February 1972 with the small Low Cost Engine in the shadow of the much larger Quiet Engine being tested in PSL No. 3. (NASA C–1972–00577)

its 600-pound-thrust limit.[60] The researchers found that operating the engine at 107 percent of its rated corrected engine speed produced 700 pounds of thrust.[61]

In 1973 the Low Cost Engine had its first realistic analysis in PSL No. 2. It was installed in the altitude chamber with a direct-connect setup and successfully operated at speeds up to Mach 1.24 and simulated altitudes of 30,000 feet. The engine was restarted several times at altitude and demonstrated the ability to perform continuously for 1 hour.[62] Follow-up tests were run at the Special Projects Laboratory the next year, and the program successfully concluded.

Lewis succeeded in creating a small turbojet with estimated production costs of around $3,000.[63] The navy flight-tested the engine at the Naval Weapons Center in California for its drone aircraft.[64]

NASA released the engine to private industry in the hope that design elements would be incorporated in future projects and reduce the overall cost of small jet aircraft.[65] Small jet and turboprop engines had become relatively common in general aviation aircraft by the late 1970s, and the center continued to pursue the field with its General Aviation Program engines.[66]

The Compass Cope

At the same time that the Low Cost Engine was being studied in PSL No. 2, another engine was being studied in PSL No. 1 for a different drone vehicle—the Teledyne Ryan Compass Cope. Development of a new generation of high-altitude, unpiloted reconnaissance aircraft to penetrate Chinese territory was instituted in 1971. The Compass Cope was similar to Lockheed's U-2 with its 81-foot wingspan, its high-altitude and long-distance flights, and its ability to take off and land on runways. The Compass Cope, however, would be deployed on missions too dangerous for a pilot.[67]

Boeing created an initial prototype in 1971, the YQM-94 B-Gulls, but Teledyne Ryan developed its own version, the YQM-98A Compass Cope Tern, the following June. Although the original Tern met all of the air force's specifications except 24-hour endurance, Teledyne Ryan decided to completely redesign the vehicle.[68]

Image 124: An early version of the Teledyne Ryan YQM-98A Compass Cope Tern with its 81-foot wingspan at Edwards Air Force Base in 1975.[69]

Image 125: Original ATF3 SN 16 engine in the PSL on 18 January 1973. This engine was later rebuilt and was available for spare parts for the Compass Cope program. The SN 17 served as the prototype for Garrett's 5,450-pound-thrust ATF3-6.[70] (NASA C–1973–00344)

The new Tern included a different engine and could perform 24-hour missions without any ground-based control. Teledyne Ryan looked to Garrett Corporation[g] for the new powerplant. Garrett had recently developed its ATF3 and TFE 731 turbofans to power several business transport aircraft.[71] The 4,050-pound-thrust ATF3 first ran in 1968, but it did not undergo flight-rate testing until January 1974. Garrett's 3,500-pound-thrust TFE 731-2 was designed in 1970 and was expeditiously flight certified in 1973.[72]

During the Tern's redesign, the air force requested the use of Lewis's PSL facility from 1971 to 1973 to compare the ATF3 and TFE 731-2 engines in altitude conditions. "Our Air Force Contract Administrator for the test managed to keep us thinking there were other competitors," Garrett engineer Jerry Steele recalled, "but I don't know if there really were any."[73]

The PSL tests would ensure that the two Garrett engines would perform throughout the aircraft's entire flight envelope. The TFE 731-2 was installed in PSL No. 1 in October 1971. The engine failed several weeks later and had to be rebuilt. When testing resumed in March 1972, the TFE 731-2 met the Tern's performance specifications but was unable to generate any additional thrust.[74]

The ATF3 tests began in early 1973. Jerry Steele, John Huber, and Jack White, respectively, served as Garrett's ATF3 test engineer, performance engineer, and controls engineer for the PSL tests.[75]

Garrett had produced two versions of the ATF3 specifically for the Tern vehicle: the SN 16 and the SN 17. The 4,050-pound-thrust SN 16 was installed in PSL No. 1 and run at a pressure-simulated 60,000 feet in January 1973. The engine performed well at that altitude, but the engineers found that the engine would

---

[g]Honeywell International, Inc., today.

stall when the power was reduced. "Thus it could go up but not come down," explained Steele.[76]

Modifications were made to the turbine nozzle, inlet guide vanes, fuel control schedules, and inlet compressor bleed system. The updated SN 16 was successfully retested at altitudes throughout the flight envelope.[77]

The SN 17 engine was then analyzed. It easily covered the proposed flight envelope, and the engineers were able use the SN 17's Electronic Engine Control. It was the first complete locked-throttle climb of any turbofan.[78]

As the program was nearing completion, the Garrett engineers decided to try to maximize the SN 17's capabilities on the last night of testing. Huber recalled the PSL operations engineer claiming that it was the first time that an engine was still generating thrust at the chamber's maximum 100,000-foot ceiling. "The inlet pressure and airflow were so low," explained Steele, "that the heavily insulated facility ducts could not prevent the outer part of the airflow next to the duct walls from heating up, causing a severe inlet temperature gradient and a very high average inlet temperature."[79]

The tests were considered to be successful, and the air force selected the ATF3 for the program. Both ATF3 engines had demonstrated the ability to cover the Compass Cope's flight envelope and had outperformed the TFE 731. In the end, the SN 16 ATF3 was chosen over the SN 17 because there was enough existing hardware to create three of the engines.[80]

Huber recalled, "After the tests were over, the Air Force Contracts Officer told us ours was the only engine in the competition to reach the minimum Mach number, maximum altitude and still have any thrust."[81] The ATF3 also had the built-in design advantage of expelling the hot exhaust through eight ejectors near the fan flow. The gases were comparably cool and quiet as they exited the nozzle. This was a valuable asset for a stealth aircraft.[82]

Flight testing of the Tern with the ATF3 engine began in August 1974. The aircraft could cruise at Mach 0.6 at 55,000 feet for more than 24 hours.[83] That November the aircraft set an endurance record for an unpiloted, unrefueled aircraft: over 28 hours. According to John Evans, a postflight fuel inspection revealed that roughly another 6 hours of propellant remained in the tanks.[84]

Nonetheless, the Boeing version was selected as the production aircraft the following year. Teledyne Ryan had outperformed Boeing in every category except cost. The program was canceled in July 1977 before the legal arguments between Teledyne Ryan and Boeing were resolved.[85,86]

The TFE 731 became the dominant engine in the midsized business jet market between 1973 and 1977. It had almost no competition in its size and power. In addition, its low noise levels and fuel efficiency coincided perfectly with the times.[87] The ATF3 went on to a successful career powering the Dassault Falcon business jets.

Hazards Exposed

It was midnight, and the third-shift crew had just arrived on duty at PSL's Equipment Building. The air-handling system was running to support a combustor rig test in the Engine Research Building. The crew was shorthanded, and there was some confusion as the

Image 126: An overpressurization in the exhausters in PSL's Equipment Building caused a serious explosion. This 7 April 1971 photograph shows windows blown out on the exterior of the building. (NASA C–1971–01269)

Image 127: Overall interior view on 7 April 1971 of damage from the PSL Equipment Building explosion. Note the broken windows in the far wall, the hole in the roof, and the raised heater at the floor level surrounded by debris. (NASA C–1971–01272)

Pursuit of Power

Image 128: Close-up of the 96-inch-diameter header that failed in the explosion. (NASA C–1971–01283)

foreman sought to staff all the required stations. In the few minutes of disorder, a valve between the exhausters and the exhaust stack was mistakenly sealed.

The sound of the overpressurization of the exhausters sent the operators scrambling to reduce the load. At 12:15 a.m. two men in the control room saw a large cloud of dust appear and immediately sought shelter on the floor as an explosion ripped through the building.[88]

The men were uninjured, but two others, 150 feet away from the building, were wounded by falling debris. The rupture of a large inverted dished head in the basement had destroyed the 6-inch-thick reinforced-concrete floor above it, damaged I-beam supports, and tore a 30-foot hole in the roof 35 feet above the floor. There was extensive damage to the equipment, piping, and windows.[89]

An Accident Investigation Board was established the following morning. One week later, after several meetings and discussions, the board presented its findings to Center Director Bruce Lundin. The investigators found that an operator had inadvertently closed a 72-inch-diameter butterfly valve, which caused the exhausters to begin perform as compressors. This soon caused the pressure in the line to increase until the large valve gave out.[90]

As a result, the PSL system was modified to incorporate several large check valves. The accident, in conjunction with another overpressurization incident several months earlier, led to NASA's implementation of a Recertification Program for test engineers.[91]

Endnotes for Chapter 6

1. Abe Silverstein interview, Cleveland, OH, conducted by John Mauer, 10 March 1989, Glenn History Collection, Oral History Collection, Cleveland, OH.
2. Virginia Dawson, *Engines and Innovation: Lewis Laboratory and American Propulsion Technology* (Washington, DC: NASA SP–4306, 1991), chap. 9.
3. Kyrill von Gersdorff, "Aeroengines—Altitude Test Beds," *Aeronautical Research in Germany: From Lilienthal Until Today*, 147, Ernst-Heinrich Hirschel, Horst Prem, and Gero Madelung, editors (Berlin: Springer-Verlag, 2004), p. 221.
4. "High-Speed Research—The Tu-144LL A Supersonic Flying Laboratory," FS–1996–09–18-LaRC (September 1996), *http://www.nasa.gov/centers/langley/news/factsheets/TU-144.html* (accessed 23 May 2011).
5. Abe Silverstein, *Progress in Aircraft Gas Turbine Engine Development* (Washington, DC: NASA TM X–52240, 1966).
6. Neal Wingenfeld interview, Cleveland, OH, 15 September 2008, NASA Glenn History Collection, Oral History Collection, Cleveland, OH.
7. Howard Wine interview, Cleveland, OH, 4 September 2005, NASA Glenn History Collection, Oral History Collection, Cleveland, OH.
8. "Their Testing Talent Advances Jet Engines," *Lewis News* (9 October 1970).
9. Leon Wenzel to the Record, "Tip-Treated J-85 Program," 31 January 1973, NASA Glenn History Collection, Test Facilities Collection, Cleveland, OH.
10. James DeRaimo, "Flame Spreader for Cell #2 at PSL," 12 August 1968, NASA Glenn History Collection, Test Facilities Collection, Cleveland, OH.

11. Fred Wilcox to the Record, "Definition of F-106 Research Program and Results of Study of Accuracy," 8 December 1967, NASA Glenn History Collection, Test Facilities Collection, Cleveland, OH.
12. "Jet Aircraft Noise Reduction," Annual Inspection at NACA Lewis Flight Propulsion Laboratory, 7–10 October 1957, NASA Glenn History Collection, Cleveland, OH.
13. Paul Colarusso to Procurement Division, "Non-Competitive Procurement for PF 442436, Design and Fabrication of Modifications to a General Electric J-85 Jet Engine to Provide Blade Tip Treatment," 7 April 1971, Glenn History Collection, Test Facilities Collection, Cleveland, OH.
14. General Electric Company, "Model J85" (2011), http://www.geae.com/engines/military/j85/index.html (accessed June 2011).
15. M. H. Campbell to Fred Wilcox, "PSL Testing of J85-13 Engine," 6 June 1967, Glenn History Collection, Test Facilities Collection, Cleveland, OH.
16. Fred Wilcox, "F106B Retirement, Comments on Its Tour of Duty at Lewis," 17 May 1991, NASA Glenn History Collection, Test Facilities Collection, Cleveland, OH.
17. Nick Samanich and Sidney Huntley, *Thrust and Pumping Characteristics of Cylindrical Ejectors Using Afterburning Turbojet Gas Generator* (Washington, DC: NASA TM X–52565, 1969).
18. "PSL Calibration Program" notes, NASA Glenn History Collection, Test Facilities Collection, Cleveland, OH.
19. W. T. Latto, Jr., "In-Flight Thrust Calculation for the J85-13 Engine," 8 February 1967, Glenn History Collection, Test Facilities Collection, Cleveland, OH.
20. John Smith, Chi Young, and Robert Antl, *Experimental Techniques for Evaluating Steady-State Jet Engine Performance in an Altitude Facility* (Washington, DC: NASA TM X–2398, 1971).
21. Smith, Young, and Antl, *Experimental Techniques for Evaluating Steady-State Jet Engine Performance.*
22. Charles Tracy, "SST Concepts are Tested Here," *The Cleveland Press*, 7 May 1969.
23. Samanich and Huntley, *Thrust and Pumping Characteristics of Cylindrical Ejectors.*
24. "Exhaust Nozzle Key to Quieter Jets," *Lewis News* (11 September 1970).
25. "Exhaust Nozzle Key to Quieter Jets."
26. "Center Recounts 1970 Progress," *Lewis News* (29 January 1971).
27. *Astronautics and Aeronautics, 1971* (Washington, DC: NASA SP–4016, 1972).
28. Fred Wilcox, "F106B Retirement."
29. "Final Flight," *Lewis News* (4 February 1977).
30. John McAulay and Mahmood Abdelwahab, *Experimental Evaluation of a TF-30-P-3 Turbofan Engine in an Altitude Facility: Afterburner Performance and Engine-Afterburner Operating Limits* (Washington, DC: NASA TN D–6839, 1972).
31. Edwin Graber and Willis Braithwaite, *Summary of Recent Investigations of Inlet Flow Distortion Effect on Engine Stability* (Washington, DC: NASA TM X–71805, 1974).
32. Ross Willoh, et al., "Engine Systems Technology," *Aeronautical Propulsion* (Washington, DC: NASA SP–381, 1954).
33. Willis Braithwaite and William Vollmar, *Performance and Stall Limits of a YTF30-P-1 Turbofan Engine with Uniform Inlet Flow* (Washington, DC: NASA TM–1803, 1969).
34. Roger Werner, Mahmood Abdelwahah, and Willis Braithwaite, *Performance and Stall Limits of an Afterburner-Equipped Turbofan Engine With and Without Engine Flow* (Washington, DC: NASA TM X–1947, 1970).
35. Willis Braithwaite, John Dicus, and John Moss, *Evaluation With a Turbofan Engine of Air Jets as a Steady-State Inlet Flow Distortion Device* (Washington, DC: NASA TM X–1955, 1970).
36. Braithwaite, Dicus, and Moss, *Evaluation With a Turbofan Engine of Air Jets.*
37. Leon Wenzel, *Experimental Investigation of the Effects of Pulse Pressure Distortions Imposed on the Inlet of a Turbofan* (Washington, DC: NASA TM X–1928, 1969).
38. McAulay and Abdelwahab, *Experimental Evaluation of a TF-30-P-3 Turbofan Engine.*
39. McAulay and Abdelwahab, *Experimental Evaluation of a TF-30-P-3 Turbofan Engine.*
40. Thomas Biesiadny, et al., *Summary of Investigations of Engine Response to Distorted Inlet Conditions* (Washington DC: NASA TM–87317, 1986).
41. *Experimental Evaluation of the TF30-P-3 Turbofan Engine in an Altitude Facility: Effect of Steady-State Temperature Distortion* (Washington, DC: NASA TM X–2921, 1973).
42. Willis Braithwaite and Ronald Soeder, *Combined Pressure and Temperature Distortion Effects on Internal Flow Through a Turbofan Engine* (Washington, DC: NASA TM–79136, 1979).
43. "Lewis Recounts Year's Progress, Looks Ahead," *Lewis News* (15 January 1971).
44. Edward Milner and Leon Wenzel, *Performance of a J85-13 Compressor With Clean and Distorted Inlet Flow* (Washington, DC: NASA TM X–3304, 1975).

45. Leon Wenzel, John Moss, and Charles Mehalic, *Effect of Casing Treatment on Performance of a Multistage Compressor* (Washington, DC: NASA TM X–3175, 1975).
46. Colarusso to Procurement Division, 1971.
47. J.S. Shannon to Wayne Park, "Transmittal of Technical Memo No. 73–240," 23 March 1973, Glenn History Collection, Test Facilities Collection, Cleveland, OH.
48. John Moss, *Effect of Slotted Casing Treatment on Performance of a Multistage Compressor* (Washington, DC: NASA TM X–3350, 1976).
49. Wenzel to the Record, 1973.
50. Shannon to Park, 1973.
51. Moss, *Effect of Slotted Casing Treatment*.
52. Wenzel, Moss, and Mehalic, *Effect of Casing Treatment*.
53. Robert Cummings and Harold Gold, *Concepts for Cost Reduction on Turbine Engines for General Aviation* (Washington, DC: NASA TM X–52951, 1971).
54. M. R. Barber and Jack Fischel, "General Aviation: The Seventies and Beyond," *Vehicle Technology for Civil Aviation* (1971): 317–332.
55. Robert Cummings, *Experience With Low Cost Jet Engines* (Washington, DC: NASA TM–X–38085, 1972).
56. "Center Tests Small Low Cost Jet Engine," *Lewis News* (24 March 1972).
57. "Center Tests Small Low Cost Jet Engine."
58. Cummings, *Experience With Low Cost Jet Engines*.
59. Robert Dengler and Lawrence Macioce, *Small, Low-Cost, Expendable Turbojet Engine—II Performance Characteristics* (Washington DC: NASA TM X–3463, 1976).
60. Cummings, *Experience with Low Cost Jet Engines*.
61. Dengler and Macioce, *Small, Low-Cost, Expendable Turbojet Engine—II*.
62. Dengler and Macioce, *Small, Low-Cost, Expendable Turbojet Engine—II*.
63. Cummings and Gold, *Concepts for Cost Reduction on Turbine Engines*.
64. Cummings, *Experience with Low Cost Jet Engines*.
65. "Center Tests Small Low Cost Jet Engine."
66. Richard Leyes and William Fleming, *The History of North American Small Gas Turbine Aircraft Engines* (Reston, VA, AIAA and the Smithsonian Institution, 1999).
67. Bill Yenne, *Attack of the Drones: A History of Unmanned Aerial Combat* (St. Paul, MN: Zenith Press, 2004).
68. John C. Evans, "Teledyne-Ryan Compass Cope YQM-98A R-Tern," The Garrett AiResearch ATF3 Online Museum (2011), http://web.me.com/jcefamily/ATF3/Ryan_Compass_Cope_2.html (accessed 15 May 2011).
69. Evans, "Teledyne-Ryan Compass Cope YQM-98A R-Tern."
70. Jerry Steele to Robert Arrighi, "Re: Altitude Testing," 23 May 2011, NASA Glenn History Collection, Test Facilities Collection.
71. Ken Fulton, "Competition Among the 5000 Pound Fans," *Flight International* (17 January 1976).
72. "Turbine Engines of the World," *Flight International* (2 January 1975).
73. Steele to Arrighi, 2011.
74. Evans, "Teledyne-Ryan Compass Cope YQM-98A R-Tern."
75. Evans, "Teledyne-Ryan Compass Cope YQM-98A R-Tern."
76. Steele to Arrighi, 2011.
77. Steele to Arrighi, 2011.
78. Evans, "Teledyne-Ryan Compass Cope YQM-98A R-Tern."
79. Steele to Arrighi, 2011.
80. Steele to Arrighi, 2011.
81. Evans, "Teledyne-Ryan Compass Cope YQM-98A R-Tern."
82. "Powerplants on Show," *Flight International* (12 June 1975).
83. Yenne, *Attack of the Drones*.
84. Evans, "Teledyne-Ryan Compass Cope YQM-98A R-Tern."
85. John Evans to Robert Arrighi, "Re: Altitude Testing," 31 May 2011.
86. Yenne, *Attack of the Drones*.
87. Leyes and Fleming, *The History of North American Small Gas Turbine Aircraft Engines*.
88. Jack Esgar, "Narrative Factual Description of Mishap," 1971, NASA Glenn History Collection, Test Facilities Collection, Cleveland, OH.
89. Jack Esgar, "NASA Industrial Mishap Report," 27 April 1971, NASA Glenn History Collection, Test Facilities Collection, Cleveland, OH.
90. Esgar, "NASA Industrial Mishap Report," 1971.
91. Vincent Verhoff, "Pressure System Safety: 1971 Bldg. 64 Explosion," 22 June 2005, NASA Glenn History Collection, Test Facilities Collection, Cleveland, OH.

Image 129: Installation of one of the two new 24-foot-diameter test chambers at the PSL No. 3 and 4 facility on 1 August 1969. (NASA C–1969–02672)

· 7 ·
The Third Step

The NASA Lewis Research Center's return to aeronautics in 1967 breathed new life into Propulsion Systems Laboratory (PSL) No. 1 and 2, but it also portended its eventual demise. Twenty years after planning began for the original PSL test chambers, the center began preparing to add two additional and more powerful altitude chambers. The expansion was a key component of the center's return to aeronautics. The original PSL chambers had been an improvement on the Four Burner Area, and the new chambers, PSL No. 3 and 4, would be an improvement on PSL No. 1 and 2. Bob Walker explained, "[It] was kind of like three steps.... The engines got bigger than the facility."[1]

The new 40-foot-long, 24-foot-diameter test sections were designed to handle engines twice as powerful as any available at the time and to test larger engines with higher temperature turbines. New compressors and exhausters, which originally could produce speeds up to Mach 3 and altitudes up to 70,000 feet, were added to the existing PSL Equipment Building.[2] This capability was expanded over the years.

Unlike PSL No. 1 and 2, the new chambers would vent their exhaust through a shared 30-foot-long, 50-foot-diameter cooler with a large davit valve to seal off the chamber not being used. This gave the new facility a Y-shape. The cooling system could reduce the temperature of the engine exhaust from 3,500°F to 150°F.[3] The hot exhaust gas passed through 17-foot-diameter water ducts then into the larger 50-foot-diameter cooler. The air was cooled as it passed through 2,700 water-filled tubes. Three banks of nozzles discharged 50 gallons of water per second to further cool the exhaust.[4]

A new heater building was built to house three heat exchangers that increased inlet air temperatures to 1,200°F for certain tests. The 165-pounds-per-square-inch air was generated by three Pratt & Whitney J57 jet engines.[5]

Another Giant Emerges
The PSL site was originally a ravine that had been filled in the 1920s to build the Cleveland Municipal Airport. After the NACA built its lab in 1942, the grassy rectangular area became known as Wright Park. The PSL No. 1 and 2 facility and its Equipment Building took over the northern half of the park in the late 1940s. Now, the PSL No. 3 and 4 facility would occupy the remainder of the area.

Excavation of the 35-foot-deep, 200-foot-wide, and 500-foot-long hole for the PSL No. 3 and 4 facility was completed in October 1967. The fill dirt was removed from the site, and pilings for the new building were extended down into the bedrock.[6]

Gilmore-Olson Construction had erected the building's 700-ton steel skeleton by the end of August 1968. The installation of the electrical, plumbing, and other infrastructure would continue throughout the next year.[7] In September 1968 the Pittsburgh-Des Moines Steel Company was hired to manufacture and install the two new test chambers and associated infrastructure.[8] Work commenced on the electrical, combustion air system, and cooling water piping in September 1970.[9]

The PSL No. 3 and 4 facility was completed in late 1972, and testing began in early 1973.[10] Center Director Bruce Lundin held a recognition ceremony for the 230 NASA Lewis personnel who brought the new facility to fruition. The various division chiefs spoke, emphasizing the coordination and teamwork required. At the end, Lundin formally dedicated PSL No. 3 and 4.[11]

Pursuit of Power

Image 130: The large open area in the middle of the NACA's Cleveland lab, seen here on 10 September 1945, was known as Wright Park. It was reserved for future expansion. (NASA C–1945–13061)

Image 131: Wright Park seen on 29 April 1957 with the PSL No. 1 and 2 facility toward the north end and the Equipment Building near the center. (NASA C–1957–44883)

Image 132: The area was completely built up by this 16 October 1987 photograph. The PSL No. 1 and 2 and PSL No. 3 and 4 facilities occupy the former park area. (NASA C–1987–09393)

Image 133: On 26 June 1967 the grassy area between the PSL No. 1 and 2 facility and the cafeteria building was bulldozed for the new PSL No. 3 and 4 facility. (NASA C–1967–02211)

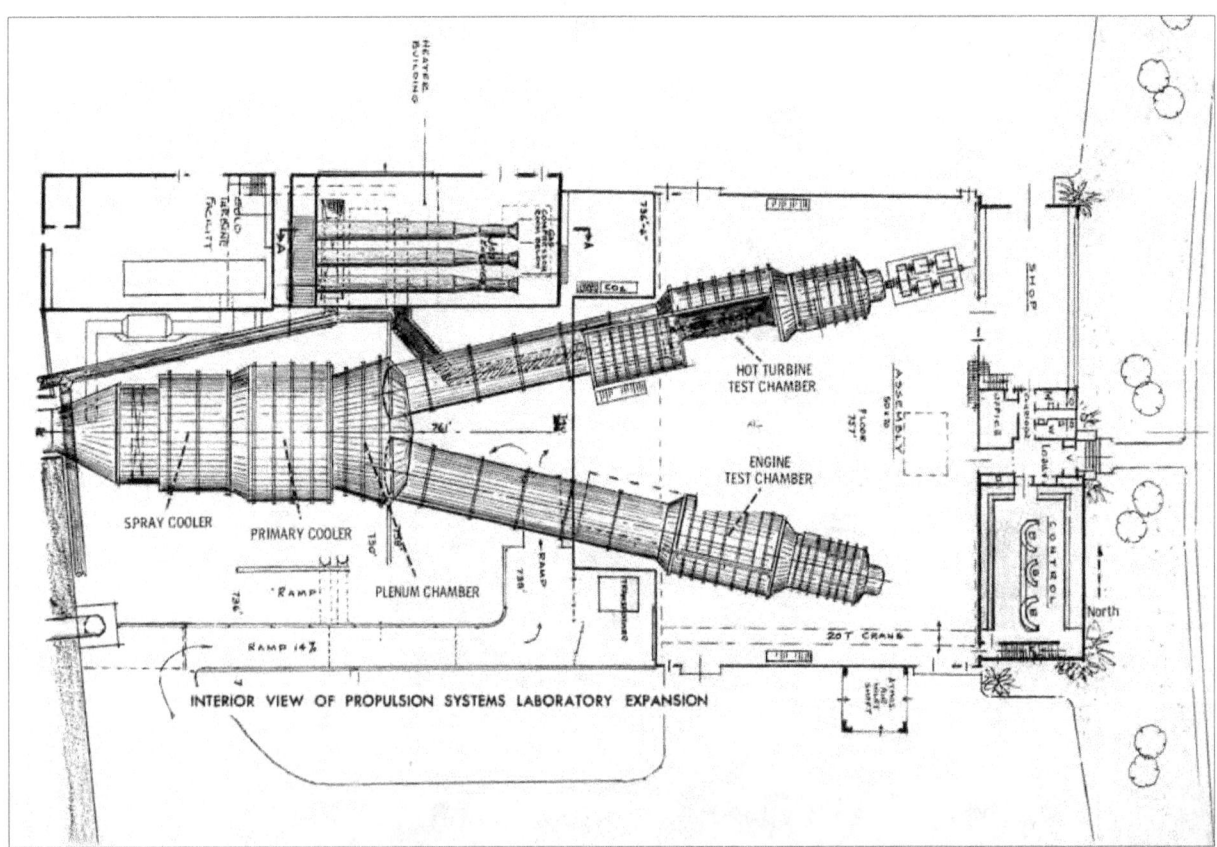

Image 134: This diagram shows the design of the new PSL No. 3 and 4 facility. The two test chambers to the right shared the common cooler at the left. The three J57 engines that supplied combustion air are near the top. (NASA Drawing)[12]

Pursuit of Power

Image 135: Construction of the massive primary cooler for the new PSL No. 3 and 4 facility on 17 November 1967. The Engine Test Building is in the background to the right. (NASA C–1969–03898)

Image 136: In this 1 June 1970 photograph, workers are placing a valve that will lead from the test chamber to the cooler. (NASA C–1970–01642)

Image 137: Construction of the two new 24-foot-diameter altitude chambers, PSL No. 3 and 4, on 6 May 1970. (NASA C–1970–01312)

Image 138: The PSL No. 3 and 4 facility seen from the air on 22 September 2005. The two altitude test chambers are located in the building to the right. The hot exhaust gas was expelled through the common plenum at the left. The Equipment Building with the exhausters and compressors is located out of sight to the left. (NASA C–2006–01485)

Image 139: The PSL No. 1 and 2 control room shown in this 2 July 1970 photograph was frequently updated, and the control panels were rearranged periodically. The manometers had been replaced with electronic units by the early 1960s. A number of television consoles were installed so that the test engineers could view the engine in the chamber during the test. Temporary data-recording equipment was installed for certain tests and later removed. (NASA C–1970–01743)

Image 140: In 1955 a fourth line of exhausters was added. The total inlet volume of the four-stage exhausters was 1.65 million cubic feet of gas per minute. The exhausters were continually improved and upgraded over the years, and they remain in operation today.[13] This photograph shows the new exhausters and compressors as they appeared on 1 April 1974. (NASA C–1974–01139)

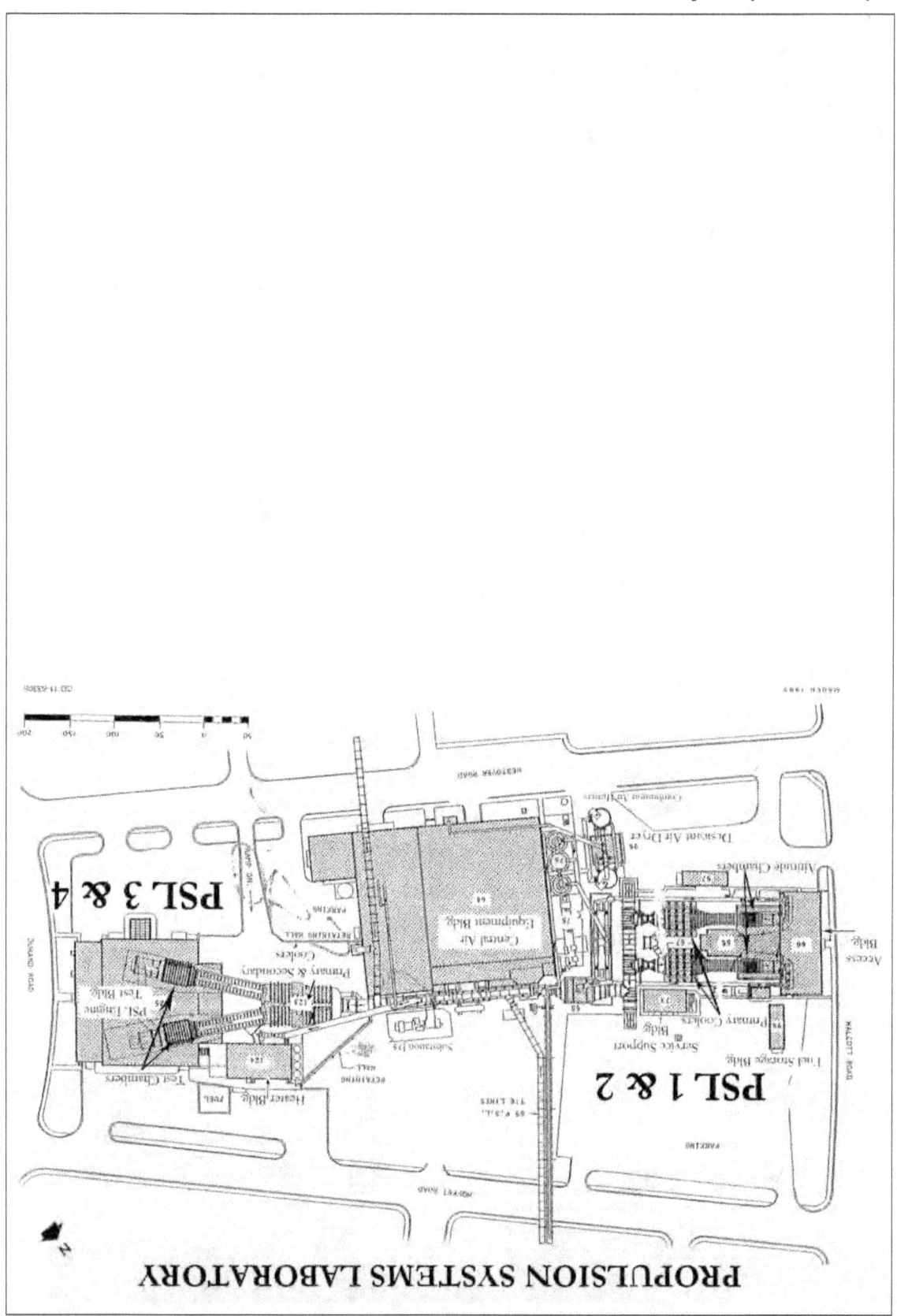

Image 141: PSL complex. (NASA CD–11–83305, adapted from a NASA report drawing).[14]

Pursuit of Power

Image 142: *The newly expanded PSL complex viewed from the southeast on 1 October 1974. The PSL No. 3 and 4 facility is in the foreground, the Equipment Building is near the center, and the PSL No. 1 and 2 facility is at the rear. (NASA C–1974–03577)*

The Lewis aeronautics program was in full swing when PSL No. 3 and 4 became operational in 1973. The center initially planned to terminate operations in PSL No. 1 and 2 when the new test chambers came online. PSL No. 1 and 2 continued operating, however, to meet the high demand for testing aircraft engines in altitude conditions. The Garrett and Low Cost Engine studies were under way, and the Airbreathing Engines Division was about to embark on several long-term, multifaceted engine studies that would use all four test chambers.

Flutter

Lewis had been investigating the effects of airflow distortions on engine inlets and compressors since the center's return to aeronautics in 1967. The nation's early jet engines were sturdy pieces of equipment that were relatively impervious to the effects of airflow distortions. Axial-flow compressor engines grew in power and sophistication during the 1950s and 1960s, increasing the number of compressor stages and incorporating dual-spool configurations. These complex powerplants brought with them a new set of operating problems, including flutter.

In order to improve performance, the stator blades were made thinner and thinner. This not only augmented the engine's capabilities under normal operating situations but resulted in a propensity to stall from flutter during abnormal conditions.

Flutter is the self-induced vibration of the compressor stator blades due to irregular airflow or distortions. The vibrations can weaken or damage the stators, which will eventually inhibit the engine's performance.

Eradicating flutter from the entire flight envelope was time consuming and expensive. It could take a year or two to redesign and retest an engine in which flutter was found. The military grappled with this issue while developing several of its engines.[15]

Full-Scale Engine Programs

In the mid-1970s, Lewis and the U.S. Air Force collaborated on two broad programs that studied a variety of design problems on full-scale engines. The first, the Full-Scale Engine Research (FSER) program, utilized surplus air force engines as testbeds for a variety of research purposes, including flutter, inlet distortion, and electronic controls. The goal was to produce technological achievements, not to resolve hardware problems on specific engines. The data were aggregated so that they could be used for future engine development efforts.[16]

The second program, the Aeroelasticity of Turbine Engines, included several projects aimed at improving compressor blade design and analysis. A better understanding of flutter was expected to lead to flutter-proof engine designs and to prevent the eventual development delays and costs.[17]

Image 143: *The F-16 Falcon was powered by the 27,000-pound-thrust F100-PW-200 turbofan. It flew more sorties than any other aircraft during Operation Desert Storm.*[18] *(2003, U.S. Air Force)*

The Aeroelasticity of Turbine Engines program used computer simulations to create analytical models but required full-scale engine testing to validate the codes. The Fluid System Components Division's Fan and Compressor Branch coordinated 9 projects at Lewis and contracted for another 19. Together the projects formed a comprehensive collection of data to assist engine designers in building their own analytical models.[19]

During the validation tests, the fully instrumented engine was installed in the PSL and operated at conditions where the problems previously occurred. The researchers then destabilized the airflow using screens and mapped the engine performance. Several engines were tested during the 1970s using this technique. From the tests, the researchers hoped to determine the effects of temperature and pressure on flutter.[20] Data on stall flutter, choke flutter, and system-mode instability were compiled into a repository to create and validate computer models.[21]

In the late 1970s, engineers from the Airbreathing Engines Division's Engine Research Branch studied two air force engines for the FSER in the PSL—the Pratt & Whitney F100 and the General Electric (GE) J85-21.

Pratt & Whitney F100

Pratt & Whitney's 23,770-pound-thrust F100 turbofan was developed in the mid-1960s almost concurrently with their TF30, and it powered modern fighters such as the North American X-15 Eagle and the Lockheed Martin X-16 Falcon.[22] In 1972 the air force began having problems with flutter in the engine, so they selected the F100 for the first FSER study in the PSL. The air force supplied Lewis with an early prototype of the engine, a YF100.[23]

PSL No. 1 had to be renovated to accommodate the 10-stage turbofan. Because the F100 was similar to the TF30, however, PSL mechanics were able to adapt and reuse much of the existing test hardware.[24] Pratt & Whitney technicians traveled to Lewis to assist with the instrumentation and to brief the researchers on the engine's flow instabilities.[25] Air force personnel made regular trips from Dayton to oversee and coordinate the Lewis testing.

Image 144: Technicians in PSL No. 1 connect wiring on an F100 engine on 14 February 1978. The F100 had a very high thrust-to-weight ratio, and its modular construction allowed easy modifications. (NASA C–1978–00476)

The flutter investigations began in the fall of 1974. A major component of the testing was mapping the airflow through the engines to identify flutter. The researchers purposely induced flutter as the engine neared its stall limits and then mapped the flutter envelope. The tests used improved instrumentation and optical devices to measure flutter.[26]

Lewis researchers later analyzed the collected flutter data, and several hypotheses were offered to explain the phenomenon.[27] Although breakthroughs to a completely flutter-free engine did not occur, several improved design techniques were developed. A 1981 Ad Hoc Aeronautics Assessment Committee concluded, "This prompts confidence in finding solutions for flutter in other operating regimes." The committee recommended expanding the aeroelasticity program to include forced vibration: "This program needs to be planned as a long-term fundamental effort."[28]

J85-21

A GE J85-21 turbojet, a 5,000-pound-thrust variant of the J85-13, was obtained from the air force in early 1975 for the FSER program. Lewis technicians began installing the J85-21 in PSL No. 2 during the fall after GE updated the engine to meet its current production standards. The engine was used for two series of investigations: internal compressor aerodynamics and mechanical instability, or flutter.[29]

During one study, Lewis researcher Roger Werner examined the effect of instrumentation on the airflow through the compressor. Werner analyzed the flow distortions caused by different combinations of rakes and vanes. A pressure rake failed during an early test run, resulting in disassembly of the engine. The tests resumed without the rakes in the compressor stages. Werner concluded that the instrumentation produced only minor distortions.[30]

Image 145: The J85-21 was a relatively small engine in comparison to the large TF30 and F100 turbofans. It is seen here on 22 March 1974 in PSL No. 2. (NASA C–1974–01012)

The researchers focused on two types of stall flutter, choke flutter, and system-mode instability. The distortion variations differed from each other, and the researchers assembled a collection of data from each type of instability.[31] The tests were interrupted, however, to perform inlet tests for the J85-21's applications.

Highly Maneuverable Aircraft Technology

Engineers at the NASA Ames Research Center, NASA Dryden Flight Research Center, and Rockwell International devised two subscale Highly Maneuverable Aircraft Technology (HiMAT) vehicles in the mid-1970s as safe post-wind-tunnel-test vehicles. The HiMAT vehicles would be used to study the behavior of fighter aircraft in the transonic realm to expedite the transition from the design phase to flight testing. These unpiloted vehicles could use new design concepts that might be too risky for a piloted vehicle.[32]

Rockwell took particular care to provide backup systems for each component while constructing the two vehicles for NASA.[33] The HiMAT aircraft were flown sequentially, the first between mid-1979 and 1981 and the second from mid-1981 to early 1983.

One aircraft focused on expanding the operating envelope and the other on data collection.[34] Specifically, they sought to perform an 8-G maneuver at Mach 0.9 and 25,000 feet.[35] Contemporary fighters had only half of that capability.[36] The aircraft used sophisticated technologies such as advanced aerodynamics, composite materials, digital integrated propulsion control, and digital fly-by-wire control systems.[37]

The HiMAT was fairly small and launched at 45,000 feet from underneath a B-52. The GE 5,000-pound-thrust J85-21 turbojet provided the HiMAT propulsion. The engine's standard hydromechanical propulsion controls were replaced with a multimode digital system. NASA pilots controlled the vehicle from a realistic cockpit in the ground station.[38]

Image 146: This photograph of one of the two HiMAT vehicles demonstrates its 0.44-scale size. (NASA C–1980–03906)

The J85-21 aerodynamics test program in the PSL was interrupted so that the engine could be used in a stall investigation for the HiMAT program. Researchers worried that distortion from the J85-21's short turning inlet would stall or hinder the HiMAT's performance. In late 1977 Lewis's Leo Burkardt and the air force's George Bobula studied the engine in PSL No. 2. They charted the inlet quality for various combinations of five screens.[39] An inlet model was also studied in Lewis's 8- by 6-Foot Supersonic Wind Tunnel in 1979.

The two HiMAT aircraft performed 11 hours of flying over the course of 26 missions from mid-1979 to January 1983 at Dryden and Ames.[40] The program demonstrated advanced fighter technologies that have been used in developing many modern high-performance military aircraft. The two vehicles provided data on the use of composites, aeroelastic tailoring, close-coupled canards, and winglets. The data were used to investigate the interaction of these then-new technologies on each other.[41]

Although the HiMAT vehicles were considered to be overly complex and expensive, the program yielded a wealth of data that would validate computer-based design tools. The program also demonstrated that multisystem technologies were beneficial.[42]

Fly-by-Wire

Since the inception of turbojets in the 1940s, engineers have been simultaneously advancing both engine performance and control. Engine control systems determine the fuel required to produce the specific levels of desired thrust. The thrust must be available despite the presence of turbulence or other abnormal flight conditions.[43]

Veteran Lewis control system researchers Sanjay Garg and Link Jaw identified four phases of control system development: the inception during the 1940s, an expansion in the 1950s and 1960s, the use of electronics in the 1970s and 1980s, and a final integration in the 1990s.[44]

The center was particularly involved in developing the new electronics systems in the 1970s, and the PSL was used to verify the performance of several of these systems in simulated altitude conditions.

Early jet engines were regulated by hydromechanical fuel-control devices with separate vacuum tube controls for the afterburners. The mechanical controls often drove the overall aircraft design, and redundancy was difficult to incorporate.[45] In the late 1940s the Cleveland lab's Altitude Wind Tunnel was used to resolve vacuum tube failures on the GE J47. The researchers determined that fuel flow and engine speed could be calculated linearly against a constant time.[46]

The larger turbojets and turbofans of the 1960s led to advances in the control systems. Aircraft engines traditionally used fixed-geometry components with variable fuel flow and nozzle areas. The new engines implemented variable-shape compressor and fan blades.

The dependable hydromechanical control systems could not keep up with the increasingly complex and powerful engines. These new types of engines required more sophisticated control systems that could handle multiple parameters and additional variables while increasing the accuracy and response of the engine.[47] Digital control technology developed for the Apollo Program was slowly taken on for aircraft propulsion. The new "fly-by-wire" electronic controls were lighter and more reliable, and they allowed greater design flexibility.[48] Variable-geometry controls included new methods for managing the compressor stators, intake, and nozzle.

Researchers began investigating multivariable control designs in the mid-1950s, but they were not perfected for another decade. The frequency response method used the gain and phase margins to maintain stability. The air force teamed with NASA Dryden in 1969 to commence a long-term digital control program.[49] Dryden acquired a Vought F-8 Crusader in 1969 and replaced its mechanical controls with a computer and wiring from the Apollo digital control system. The F-8 flew the first-ever digitally controlled flight on 25 May 1972. It went on to demonstrate the feasibility of the system over 40 flights.[50]

Image 147: NASA research pilot Gary Krier beside the Dryden F-8 Digital Fly-By-Wire aircraft. He flew the first fly-by-wire flight on 25 May 1972. (NASA ECN–3091)

Remote Control

NASA researchers next sought to integrate all engine components through a single digital computer. Digital systems would facilitate the control of future multivariable engines and reduce hardware update costs. During its pursuit of the supersonic transport design, Boeing engineers realized that the industry's tradition of integrating propulsion and flight controls in a single system could not be automatically applied to new high-performance engines. A different philosophy was needed. The first step would be creating a digital control system for all of the engine's components.[51,52]

The air force initiated the Integrated Propulsion Control System (IPCS) program in March 1973 to demonstrate the digital control of the engine inlet, afterburner, and nozzle. Lewis developed a novel method to integrate the control of the inlet and engine, and Dryden led the effort to flight-test the system.[53]

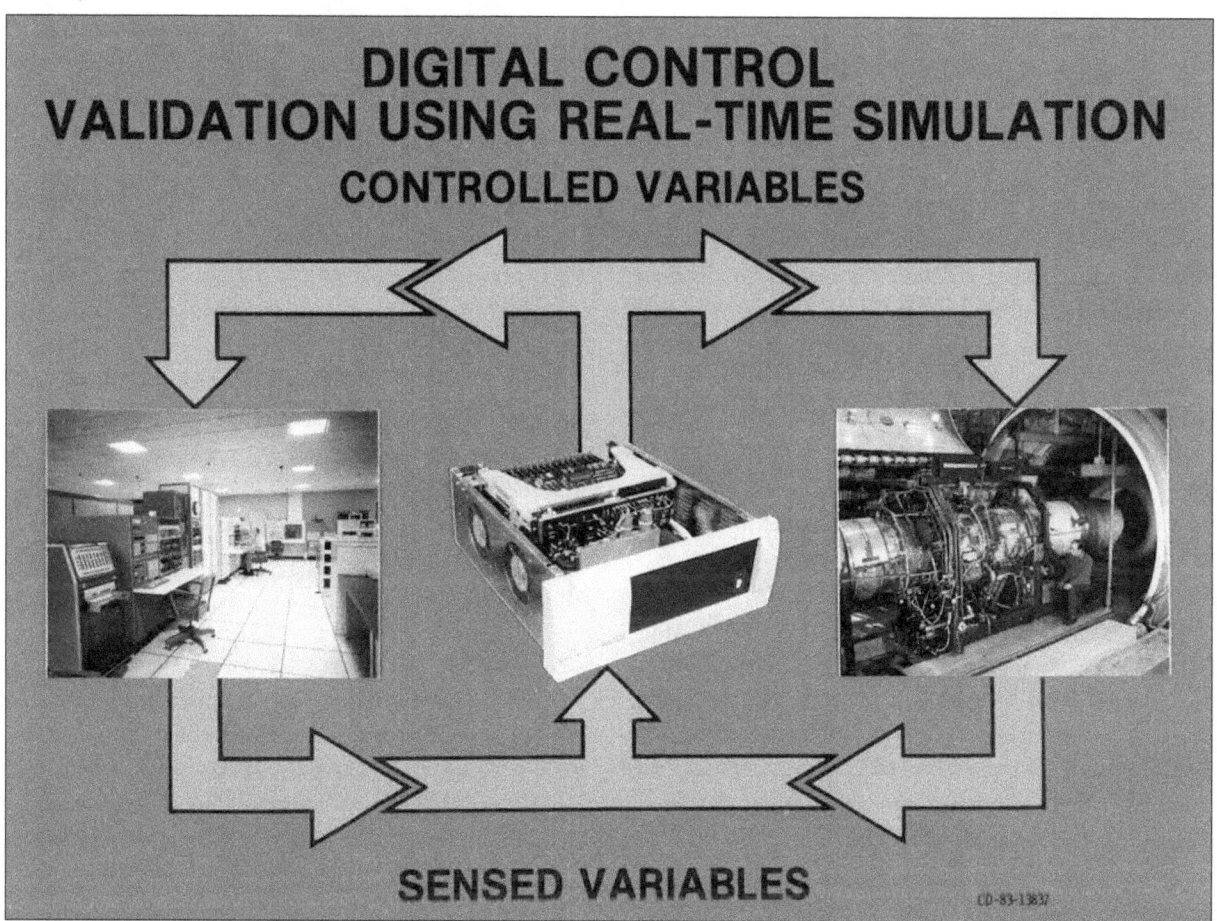

Image 148: Chart outlining the PSL validation of computer-simulated concepts such as digital engine controls. (NASA C–1983–02762)

The Lewis IPCS was installed on a Pratt & Whitney TF30-P-9 engine and tested on sea-level test stands at Honeywell and Pratt & Whitney.[54] Dryden acquired an F-111 Aardvark fighter jet in mid-1974 to flight-test the system. The IPCS was installed on one of the aircraft's two TF30 engines. As a safety precaution, the other engine remained manually controlled. According to historian Richard Hallion, there was tension between the Lewis and Dryden personnel regarding the IPCS for almost a year before the installation began in March 1975.[55]

The final engine checkout before the flight tests was a verification of the IPCS in PSL's simulated altitude conditions. From April to August 1975 the TF30-P-9 was run in PSL No. 2 at Mach 1.9 and at high-altitude low-speed flight conditions.[56] Lewis engineers used PSL data to make final adjustments to the analog inlet simulation.[57]

On 4 September 1975 the F-111 completed the first IPCS flight at Dryden. Over the next six months, 14 additional IPCS flights were successfully completed. The system was not perfect, but it did perform as efficiently as the mechanically controlled engine.[58]

Multivariable Control

Early electronic controls used closed-loop systems for a small number of variables. It was difficult for these systems to manage the many variables of modern engines, particularly vertical and/or short takeoff and landing engines. Research into multivariable control systems (MVCS) began in the early 1970s. There were two main fields: the frequency domain and the time domain. Researchers devised a linear quadratic equation in an early time-domain attempt to resolve the problem. As early as 1973 the air force began applying MVCS to the GE J85 engine.[59]

In 1975 the air force asked Lewis to build on the earlier linear quadratic regulator work and to create a realistic MVCS for the F100 engine. Pratt & Whitney provided the initial F100 computer simulations, and Systems Control, Inc., developed the computer logic for the system.[60] The Lewis researchers worked with the Systems Control engineers to develop a series of time-domain linear quadratic regulators to produce the desired transient controls. These equations would convert the pilot's input into optimal engine performance. A Lewis computer then provided the data to answer the equations. This was the MVCS.

Project manager Bruce Lehtinen said at the time, "The aim of the project was to develop and demonstrate a design procedure, not a piece of hardware."[61]

The design of the Lewis system, referred to at the time as a "smart carburetor," took only 10 months. The primary component of the MVCS was a computer with an array of control system conditions for any flight operating condition. The MVCS controlled the engine's fuel flow, exhaust nozzle, compressor bleed, and fan and compressor vanes.[62]

The MVCS was verified for nearly two years using computer simulations in Lewis's Hybrid Computer Laboratory. Then it was tested in mid-1977 on an F100 engine in the PSL.[63] The main computer in the computer lab was linked to the PSL test chamber by 1,000-foot-long underground cables. Thus, the remote MVCS could operate the F100 throughout its entire flight envelope in PSL's altitude chamber.[64]

The linear quadratic regulator proved itself applicable to flight design digital computers.[65] The digital system was more accurate than traditional mechanical controls. The system provided increased reliability, precision, responsiveness, ease of replication, and life span.[66]

Complete Control

The final phase in the evolution of digital control system technology was Full Authority Digital Engine Control (FADEC). FADECs control every aspect of an aircraft engine for maximum efficiency without pilot override. The U.S. Navy requested that Lewis develop an engine-mounted propulsion control system for fighter jets. They sought to reduce replacement costs while advancing system functionality, reliability, and performance.[67]

Lewis worked with Pratt & Whitney to develop a FADEC system for the FY-401 turbofan, a variation of the F100 originally intended for the navy's F-14 and F-15 programs. The navy dropped its contract for the F-401s in 1973, but it kept the door open for future implementation.[68]

The system included microelectronics, extensive fault tolerance, and high-speed digital communication. It was the first time that optical digital data communication had been incorporated into an engine.[69]

Pratt & Whitney performed the initial bench testing and sea-level afterburner tests of the F-401. In the spring of 1979 the engine and FADEC system were

Image 149: Dryden's F-15A was equipped with its Digital Electronic Engine Control system. In 1981 it became the first aircraft to employ a FADEC system. (NASA ECN–18899)

| COMPLETED SCHEDULES OF NASA-LEWIS WIND TUNNELS, FACILITIES AND AIRCRAFT | 1972 to 1974 | 02-13-87 | page 95 |

	1972 J F M A M J J A S O N D	1973 J F M A M J J A S O N D	1974 J F M A M J J A S O N D	OPERATIONS and RESEARCH
PSL-1				
TFE 731-2 Program	T T T T T T T T			
ATF-3 Program		I T T T T T		
TF-30 P-9 IPCS Program		I T T T T T T		
TF-30 P-1 Afterburner Program			I T T	
F-100 FX213 Fan Flutter – Phase I			I T T T T	
PSL-2				
J85-13 Inlet Distortion Rig	I T T T T T T T T T			
NASA Low Cost Engine		I T T T T T T T		
J85-13 Tip Treatment		I T T T T	I T	
J85-13 Ram Air Cooled Plug			I T T T T T T	

A = Assembly C = Conflict H = Hardware I = Installation K = Checkout M = Maintenance
O = Occupied Q = Qualification Tests R = Retrofit T = Testing ∂ = Mandatory Completion

Image 150: *Page from a schedule book showing testing in PSL No. 1 and 2 during 1972–74. (NASA)*[70]

run in PSL No. 2 at nine simulated altitudes from 7,000 to 50,000 feet.[71] Dryden acquired an F-15 Eagle afterward and installed a FADEC system on its F100 powerplant. The F-15 flew the first flight of a FADEC system in 1981. Because of the success of these tests, the air force decided to put the system into production.[72]

FADEC systems are a mainstay of current aircraft engines. They allow economical and reliable operation of the engine by receiving and instantly responding to an array of sensor inputs. FADEC systems are so prevalent that the use of mechanical systems would be nearly impossible on modern engines.[73]

Glory Days
"The 1960s were glory days of aircraft engine development," proclaimed William Hong and Paul Collopy in a 2005 paper. Strides in design techniques flourished, and almost all modern aircraft engine technology matured during this period. The nation's first turbofans emerged, influencing engine design for almost 30 years. The relationship between the air force, industry, and NASA was the closest that it had ever been.[74]

The 1960s and 1970s were also the glory days for the PSL. The use of full-scale engine models was crucial to the understanding of system integration, the perfection of technologies, and the determination of which technologies to pursue. Lewis's full-scale engine programs vetted new propulsion technologies. The PSL was NASA's only facility capable of carrying out tests of these full-scale engines in simulated flight conditions.

The full-scale programs were cresting when PSL No. 3 and 4 came online in 1972. The four PSL chambers demonstrated that they could work together in a complementary way on a single program. The new chambers were also used to study the TF30, F100, and J85 during this period.

Although the PSL was busier than ever, the 1970s were the center's bleakest years with budget and staff levels falling. As programs were cut, the transfer of engine component technology from NASA to industry nearly came to a standstill. In addition, engine development decreased from two projects each year in the 1960s to virtually none in the 1990s.[75] As PSL's most intense period of study came to an end, the future of the original test chambers began to cloud.

Endnotes for Chapter 7

1. Robert Walker interview, Cleveland, OH, 2 August 2005, NASA Glenn History Collection, Oral History Collection, Cleveland, OH.
2. "PSL Expansion Building 70% Complete Mark," *Lewis News* (25 September 1970).
3. "The Propulsion Systems Laboratory" (Cleveland, OH, NASA B–0489, 1992).
4. "$3.29 Million Contract Let for PSL Expansion," *Lewis News* (13 September 1968).
5. "PSL Expansion Building 70% Complete Mark."
6. "Work on PSL Expansion in Early Stages," *Lewis News* (18 August 1967).
7. "PSL Expands Out and Up," *Lewis News* (27 September 1968).
8. "$3.29 Million Contract Let for PSL Expansion."
9. "PSL Expansion Building 70% Complete Mark."
10. Robert Godman, "Engineering Services," *Lewis News* (29 December 1972).
11. "Awareness Group Honors PSL Team," *Lewis News* (22 February 1974).
12. "$3.29 Million Contract Let for PSL Expansion."
13. "Major Research Facilities of the Lewis Flight Propulsion Laboratory, NACA Cleveland, Ohio, Wind Tunnels—Propulsion Systems Laboratory," 19 July 1956, NASA Glenn History Collection, Cleveland, OH.
14. "Plans for Buildings and Structures," NASA Lewis Research Center, 1985.
15. Ad Hoc Aeronautics Assessment Committee, *NASA's Aeronautics Program: Systems Technology and Experimental Programs* (Washington, DC: NASA CR–164642, 1981).
16. W. J. Deskin and H. G. Hurrell, *Summary of NASA/Air Force Full Scale Engine Research Using the F100 Engine* (New York, NY: AIAA 79-1308, 1979).
17. "The Fan-Compressor Flutter Team," *Lewis News* (3 February 1978).
18. "F-16 Fighting Falcon," National Museum of the United States Air Force (2009), http://www.af.mil/information/factsheets/factsheet.asp?fsID=103 (accessed 10 April 2011).
19. "The Fan-Compressor Flutter Team."
20. Ross Willoh, et al., "Engine Systems Technology," *Aeronautical Propulsion* (Washington, DC: NASA SP–381, 1954).
21. Joseph Lubomski, *Characteristics of Aeroelastic Instabilities in Turbomachinery: NASA Full Scale Engine Test Results* (Washington, DC: NASA TM–79085, 1979).
22. "Pratt & Whitney F100-PW-220," National Museum of the United States Air Force (2009), http://www.nationalmuseum.af.mil/factsheets/factsheet.asp?id=15985 (accessed 10 April 2011).
23. Deskin and Hurrell, *Summary of NASA/Air Force Full Scale Engine Research*.
24. Walter Bishop to Chief of Engineering Design Division, "Full-Scale Engine Program for PSL Cell #2," 14 September 1973, Glenn History Collection, Test Facilities Collection, Cleveland, OH.
25. Deskin and Hurrell, *Summary of NASA/Air Force Full Scale Engine Research*.
26. "The Fan-Compressor Flutter Team."
27. Deskin and Hurrell, *Summary of NASA/Air Force Full Scale Engine Research*.
28. Ad Hoc Aeronautics Assessment Committee, *NASA's Aeronautics Program*.
29. Roger Werner, *Steady-State Performance of a J85-21 Compressor at 100 Percent of Design Speed With and Without Interstage Rake Blockage* (Washington, DC: NASA TM–81451, 1980).
30. Werner, *Steady-State Performance of a J85-21 Compressor*.
31. Lubomski, *Characteristics of Aeroelastic Instabilities in Turbomachinery*.
32. Dwain Deets, V. Michael DeAngelis, and David Lux, *HiMAT Flight Program: Test Results and Program Assessment Overview* (Washington, DC: NASA TM–86725, 1988).
33. Robert Kempel and Michael Earls, *Flight Control Systems Development and Flight Test Experience With the HiMAT Research Vehicles* (Washington, DC: NASA TP–2822, 1988).
34. Deets, DeAngelis, and Lux, *HiMAT Flight Program*.
35. Kempel and Earls, *Flight Control Systems Development and Flight Test Experience*.
36. Ken Atchison, "NASA HIMAT Craft to Test Advanced Aircraft Maneuverability," NASA News, No. 78–99, 12 July 1978.
37. Kempel and Earls, *Flight Control Systems Development and Flight Test Experience*.
38. Kempel and Earls, *Flight Control Systems Development and Flight Test Experience*.
39. George Bobula and Leo Burkardt, *Effects of Steady-State Pressure Distortion on the Stall Margin of a J85-21 Turbojet Engine* (Washington, DC: NASA TM–79123, 1979).
40. Kempel and Earls, *Flight Control Systems Development and Flight Test Experience*.

41. Dryden Flight Research Center, "Highly Maneuverable Aircraft Technology (HiMAT) Aircraft Photo Gallery Contact Sheet" (12 May 2000), *http://www.dfrc.nasa.gov/gallery/photo/HiMAT/HTML/index.html* (accessed 25 May 2011).
42. Deets, DeAngelis, and Lux, *HiMAT Flight Program*.
43. Link Jaw and Sanjay Garg, *Propulsion Control Technology Development in the United States* (Washington, DC: NASA/TM–2005-213978, 2005).
44. Jaw and Garg, *Propulsion Control Technology Development in the United States*.
45. Richard Hallion, *On the Frontier: Flight Research at Dryden, 1946–1981* (Washington, DC: NASA SP–4303, 1984).
46. Jaw and Garg, *Propulsion Control Technology Development in the United States*.
47. John Szuch, et al., *F100 Multivariable Control Synthesis Program—Evaluation of a Multivariable Control Using a Real-Time Engine Simulation* (Washington, DC: NASA TP–1056, 1977).
48. Hallion, *On the Frontier: Flight Research at Dryden*.
49. Jaw and Garg, *Propulsion Control Technology Development in the United States*.
50. Hallion, *On the Frontier: Flight Research at Dryden*.
51. L. O. Billig, J. Kniat, and R. D. Schmidt, *IPCS Implications for Future Supersonic Transport Aircraft* (Washington, DC: N76–31135 22–01, 1976).
52. NASA Dryden Flight Research Center, "Advanced Control Technology and its Potential for Future Transport Aircraft" 453–475.
53. Billig, Kniat, and Schmidt, *IPCS Implications for Future Supersonic Transport Aircraft*.
54. Billig, Kniat, and Schmidt, *IPCS Implications for Future Supersonic Transport Aircraft*.
55. Hallion, *On the Frontier: Flight Research at Dryden*.
56. Hallion, *On the Frontier: Flight Research at Dryden*.
57. Billig, Kniat, and Schmidt, *IPCS Implications for Future Supersonic Transport Aircraft*.
58. Hallion, *On the Frontier: Flight Research at Dryden*.
59. Szuch, et al., *F100 Multivariable Control Synthesis Program*.
60. *F100 Multivariable Control System: Engine Models/Design Criteria* (West Palm Beach, FL: United Technologies Corp. AFAPL–TF–7674, 1976).
61. "Progress Noted in Jet Engine Control Study," *Lewis News* (14 September 1979).
62. "Progress Noted in Jet Engine Control Study."
63. Jaw and Garg, *Propulsion Control Technology Development in the United States*.
64. "Progress Noted in Jet Engine Control Study."
65. Szuch, et al., *F100 Multivariable Control Synthesis Program*.
66. "Progress Noted in Jet Engine Control Study."
67. "International Turbine Engine Directory," *Flight International*, 7 January 1978.
68. "International Turbine Engine Directory."
69. R. W. Vizzini, T. G. Lenox, and R. J. Miller, *Full Authority Digital Electronic Control Turbofan Engine Demonstration* (Los Angeles: SAE–801199, 1980).
70. Ronald Blaha, "Completed Schedules of NASA Lewis Wind Tunnels, Facilities and Aircraft 1944–1986," February 1987, NASA Glenn History Collection, Test Facilities Collection, Cleveland, OH.
71. Vizzini, Lenox, and Miller, *Full Authority Digital Electronic Control Turbofan Engine*.
72. Jaw and Garg, *Propulsion Control Technology Development in the United States*.
73. D. Kurtz, Dr. J. W. H. Chivers, and A. A. Ned Kulite, *Sensor Requirements for Active Gas Turbine Engine Control* (Leonia, NJ: Kulite Semiconductor Products, Inc., 2001).
74. William Hong and Paul Collopy, "Technology for Jet Engines: Case Study in Science and Technology Development," *Journal of Propulsion and Power*, 21, no. 5 (September-October 2005).
75. Hong and Collopy, "Technology for Jet Engines."

Image 151: NASA Glenn demolished PSL No. 1 and 2 during the summer of 2009. The destruction is nearly complete in this 1 July 2009 photograph. Only one primary cooler remains standing. (C–2009–02016)

· 8 ·
No Tomorrow

Clouds were general all over Cleveland. A breeze gathered and blew through every part of the NASA Glenn Research Center,[h] blowing past the flags in front of the Administration Building, blowing the hair of those on their way to the cafeteria, and farther westward, blowing over the large wind tunnels and the block houses of the Rocket Lab. It was blowing, too, on the dusty wasteland where the steel carcass that had been the Propulsion Systems Laboratory (PSL) lay temporarily frozen in its final throes.

It was 1 July 2009, and NASA photographer Bridget Caswell stood amongst the dusty aftermath with her camera. She was documenting PSL's final days. Almost 59 years after her predecessors had snapped photographs of the gleaming new PSL, the facility looked as if it had been battered by the very enemy bombers that the skywatchers had vigilantly guarded against in 1952. However, it was budgets, not bombs, that brought the PSL down.

The Long Winter
NASA Lewis Research Center's funding was drastically cut in the post-Apollo years. NASA had canceled the nuclear propulsion and power programs and reallocated funds to other centers more involved with the design of the space shuttle. After large reduction-in-force actions in 1973 and 1974, Lewis's funding and staffing levels continued to decline throughout the decade. Lewis successfully moved into the fields of renewable energy, Earth resources, and energy-efficient engines. Nonetheless, Lewis continued to struggle for allocations from NASA's ever-shrinking budgets.

Center Director Bruce Lundin retired in 1977, severing Lewis's connection to its early NACA era. His replacement, outsider Eugene McCarthy, reorganized the center in 1978. The Airbreathing Engines Division was disbanded, with some of the duties being taken up by the new Propulsion Systems Division.

Staffing the test facilities continued to be a struggle. The center decided to reduce the number of technicians and mechanics to trim operating costs. This led to the consolidation of the crews for the PSL and the two large supersonic wind tunnels. In addition, these main test facilities were once again limited to overnight operations.[1] In 1978 Lewis management decided to shutter PSL No. 1 and 2 and several smaller facilities so that the 20 technicians that had been stationed there could operate other facilities.[2] The PSL made its final runs in late 1979.

Contractors were given the first floor area of the Shop and Access Building where the thermocouple and instrument departments had been located. The second floor was soon utilized as office space for engineering and maintenance staff. Temporary offices were crudely wedged in between and around the two test chambers. These makeshift office areas were used right up until 2008, but the test facility itself sat idle for 40 years.

Over the years, the countless pipes, access platforms, and other equipment began suffering from lack of maintenance. "I think they turned the whole building over to contractors because they started disassembling some of the piping to make room for offices," remembered Neal Wingenfeld.[3]

It is important to note, however, that the Equipment Building, now referred to as the "Central Air and Equipment Building," and PSL No. 3 and 4 continued to operate and be upgraded over the years. They are still critical to the center's operations.

---
[h] The center name was changed on 1 March 1999 to the NASA John H. Glenn Research Center.

Image 152: Interior of the Shop and Access Building on 22 August 2005 from above PSL No. 2. Temporary offices can be seen to the left of the chamber in the foreground. (NASA C–2008–04155)

Demolition Decision

NASA Glenn began reexamining its facilities and infrastructure in 2003. NASA Headquarters had offered the NASA centers funding to remove their unused structures. Glenn responded with a list of nine buildings that it wanted to demolish. Two of these, the Altitude Wind Tunnel and the PSL No. 1 and 2 facility, had played significant roles in advancing the nation's propulsion technology.

Although it was once a premier facility, reactivation was not an option. The piping and air systems for the original test chambers would have had to be recertified; the mechanical, electrical, and safety equipment was obsolete; and the control room and all the instrumentation had been cannibalized.[4] In addition, PSL No. 3 and 4 seemed capable of handing NASA's engine testing requirements.

Glenn was thus spending $76,000 annually to maintain an underutilized office space. NASA Headquarters agreed to the $3.17 million proposal to raze the facility, and Glenn spent the next two years creating a demolition plan and soliciting bids to do the work.[5]

Image 153: Interior of the atmospheric air vent behind the test chambers on 9 September 2008. The facility had been inactive for 30 years. (NASA C–2008–04148)

Image 154: Cannibalized control room for PSL No. 1 and 2 on 9 September 2008. The control panels, data systems, and equipment have been removed. The only indication of the room's original function is the combustion air system indicator board hanging on the far wall. (NASA C–2008–04125)

Pulling PSL Down

The demolition of a government facility requires extensive planning and recordkeeping. NASA Glenn began by creating a requirements document in September 2004 and a Statement of Work in 2005. The Ohio Historic Preservation Office was notified in May 2004 that the center was removing a significant structure. A Section 106 report was submitted to the Ohio Historic Preservation Office in July 2006 and approved in September 2007. Design services were obtained, and demolition plans were created. Bids to perform the work were solicited, and the contract was awarded in 2007.[6]

The demolition consisted of three phases: relocation of the utilities, lead paint and asbestos remediation, and the actual demolition of the facility.

The first step was the installation of perimeter fencing around the site in the spring of 2008. This was followed by the abatement process, and the removal of small amounts of mercury and lubricating oils. That September, the facility's transite walls were removed.[7]

The main demolition began in May 2009 with the removal of the Service Support Building and external pipes. Work ramped up quickly in June. The bulldozers tore into the Shop and Access Building and methodically ripped the two altitude tanks into pieces as the workers hosed down the dust.

The interior of the massive primary coolers stood exposed for the first time in 60 years as the rubble piled up around them. The cooling vanes lay in tangled piles like an industrial haystack. By August it was all over. The coolers had been knocked down, and the debris had been loaded into trucks and hauled away. Approximately 1,000 tons of steel had been removed and recycled.[8] Crews had removed the concrete foundations, graded the area, and slowly transformed the site into a parking lot and grassy area.[9]

NASA's Propulsion Systems Laboratory No. 1 and 2

Image 155: PSL wasteland on 1 July 2009. In just over two months, the wrecking crew leveled a facility that had been part of the center's landscape for nearly 60 years. (NASA C–2009–02018)

Image 156: The PSL emerging in September 1952 as the nation's premier engine testing facility. (NASA C–1952–30764)

No Tomorrow   141

Image 157: As an early sign of the impending demolition on 15 December 2008, the combustion air piping entering the Shop and Access Building has been marked for removal. (NASA C–2008–04472)

Image 158: The Service Support Building's exterior walls are removed in September 2008 as one of the first steps in the demolition of PSL No. 1 and 2. (NASA C–2008–04132)

Image 159: A steam shovel tears apart the PSL Shop and Access Building on 3 June 2009. (NASA C–2009–01660)

Image 160: Steel cooling vanes from the primary cooler lie in a pile in the foreground on 22 July 2009. A crane tears into the primary cooler in the background. (NASA C–2009–02021)

Pursuit of Power

Images 161–162: PSL No. 2 is ripped out by the demolition crew on 3 June 2009. The massive steel chambers and reinforced concrete were no match for the machine. (Top: NASA C–2009–01688; Bottom: NASA C–2009–01692)

Image 163: On 22 August 2005, Ohio State Historic Preservation Officer Lisa Adkins met with NASA Glenn officials to tour two sites scheduled for demolition—the PSL No. 1 and 2 facility and the Altitude Wind Tunnel. Adkins talks with Glenn Chief Architect Joe Morris in the PSL control room. (NASA C–2008–04153)

Historical Mitigation

The PSL No. 1 and 2 facility was considered to be eligible for, but was not listed on, the National Register of Historical Places for its contributions to aeronautics and spaceflight in the United States, particularly in the development of turbojet engines and the RL-10 rocket engine. Section 106 of the National Historic Preservation Act mandates that, for all demolitions of historical structures at federal agencies, including NASA, formal notification of the State Historic Preservation Office is required before the start of the project. The agency and the State Historic Preservation Office must reach an agreement on an appropriate level of documentation or mitigation of the facility prior to any work being performed. In addition, the public must be given an opportunity to comment on the project prior to the demolition work. This meeting was held at the center on 27 April 2006.

Glenn's Historic Preservation Officer, History Program, and Facilities Division worked with the Ohio State Historic Preservation Office to develop a plan to create a permanent documentary record of PSL No. 1 and 2 that would increase public awareness of the center's contributions to society, provide educational resources, and create a collected body of materials for future researchers.

Glenn began this process for the PSL by informing the Ohio State Historic Preservation Office of its plans. An agreement was made that permitted NASA to go forward with the demolition if they funded historic documentation of the facility.

The center undertook a broad effort to both physically document the PSL and compile the history of its construction, research, and contributions to the

Image 164: Howard Wine in one of the PSL's chambers on 16 April 1962. (NASA C–1962–60068)

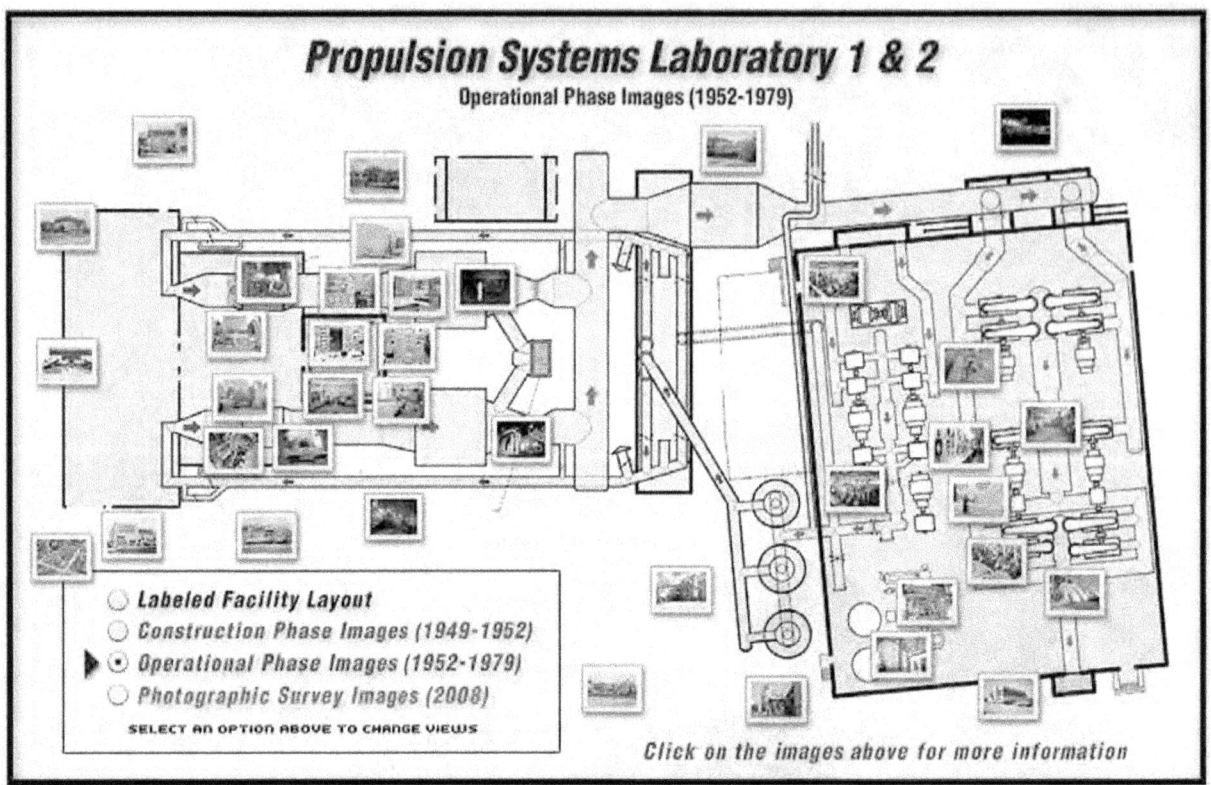

Image 165. Screen capture from interactive piece showing photographs from different areas at the PSL complex. (NASA, available at http://pslhistory.grc.nasa.gov)

nation's aerospace community. The facility was extensively photographed prior to its removal. Documents, photographs, blueprints, films, and oral histories were gathered. These materials were used to create this publication, facility drawings, a website (*http://pslhistory.grc.nasa.gov*), and an exhibit display.

PSL Legacy

PSL No. 1 and 2 served as a major component of the center's advanced propulsion legacy that began in 1942 and continues today. The facility was a technological combination of the old static sea-level test stands and the complex Altitude Wind Tunnel, which recreated actual flight conditions on a larger scale.

PSL's significance lies in the size and power of the engines it tested. When it became operational in 1952, the PSL was the nation's only facility that could run these large full-size engine systems in controlled altitude conditions. The ability to control the test environment was imperative in advancing the ever-increasing and complex turbojet systems. Today, PSL No. 1 and 2's successor, PSL No. 3 and 4, is NASA's only facility with this capability.

Much of PSL's significance and history has faded during the years since the original chambers were closed. Although the facility is now gone, it is hoped that this book and related research will restore appreciation of PSL's contributions to complex programs such as the Navaho Missile, RL-10 rocket engine, and F100 turbofan; serve as a reminder of the significant boost to the capabilities of many of the center's other facilities provided by PSL's infrastructure; and preserve the legacy of the facility operators and technicians who repeatedly brought the massive tangle of steel to life as well as that of the researchers whose investigations yielded a wealth of propulsion knowledge.

Endnotes for Chapter 8

1. James Connors, "Technical Services," *Lewis News* (4 January 1980).
2. "Answer Line," *Lewis News* (26 May 1978).
3. Howard Wine interview, Cleveland, OH, 4 September 2005, Glenn History Collection, Oral History Collection, Cleveland, OH.
4. Leslie Main, "Recordation of the Glenn Research Center Propulsion Systems Laboratory No. 1 and 2, Section 106 Check Sheets," 11 October 2004, NASA Glenn History Collection, Test Facilities Collection, Cleveland, OH.
5. "Project Requirements Document for Demolition of the Propulsion Systems Laboratory Cells 1 & 2 at Glenn Research Center. Project No. 630," 24 May 2005, NASA Glenn History Collection, Test Facilities Collection, Cleveland, OH.
6. Main, "Recordation of…Section 106 Check Sheets."
7. "Project Requirements Document for Demolition of the Propulsion Systems Laboratory."
8. Main "Recordation of…Section 106 Check Sheets."
9. "Project Requirements Document for Demolition of the Propulsion Systems Laboratory."

· Bibliographic Essay ·

Much of the information for this manuscript was gleaned from materials in the NASA Glenn History Collection. Historical correspondence and reports from our Director's office, talks given at the NACA's Triennial Inspections, the complete run of our center newspaper, and oral history transcripts provided material for a significant portion of the manuscript. The source material used by Virginia Dawson to write *Engines and Innovation: Lewis Laboratory and American Propulsion Technology* (NASA SP–4306, 1991) is archived in the NASA Glenn History Collection and contributed greatly to the early chapters of this work. In addition, many materials gathered for *Revolutionary Atmosphere: The Story of the Altitude Wind Tunnel and the Space Power Chambers* (NASA SP–2010–4319) were also applicable to this manuscript.

Glenn's Imaging Technology Center provided access to the center's extensive photograph and film collections. The photographs were used not only to highlight the text but to establish the chronology and structure of the Propulsion Systems Laboratory (PSL) story.

The massive archive of NACA and NASA technical reports in the Aeronautics and Space Database provided much of the information regarding the actual tests in the PSL. Over 65 of these reports were used for this publication.

A great deal of information was obtained from Les Main, NASA Glenn's Historic Preservation Officer. In addition, six interviews were conducted with retirees: Howard Wine and Bob Walker were familiar with the 1950s and 1960s testing; Neal Wingenfeld, John Kobak, and Larry Ross were experts on the Centaur period; and former Garrett Corporation engineers Jerry Steele and John Evans were helpful regarding the ATF3 tests. Transcripts of these interviews are available in the Glenn History Collection.

Approximately 30 secondary resources were consulted for contextual information. The most important were Dawson's previously mentioned *Engines and Innovation*, Alex Roland's *Model Research v1-2* (NASA SP–4103), Russ Murray's "The Navaho Inheritance," Richard McMullen's "History of Air Defense Weapons," Rexmond Cochrane's *Measures for Progress: A History of the National Bureau of Standards*, and Richard Hallion's *On the Frontier of Flight: Flight Research at Dryden, 1946–1981*. Others included Carl Stechman and Robert Allen's *History of Ramjet Propulsion Development at the Marquardt Company— 1944 to 1970*; Chris Gainor's, *Arrows to the Moon: Avro's Engineers and the Space Race*; Ernst-Heinrich Hirschel, Horst Prem, and Gero Madelung's *Aeronautical Research in Germany: from Lilienthal Until Today*; and the National Museum of the Air Force.

· Interview List ·

Friedman, Harold, 2 Nov. 2005 (Beachwood, OH)

Kobak, John, 1 Sept. 2009 (Cleveland, OH)

Ross, Larry, 1 Mar. 2007 (Cleveland, OH)

Walker, Robert, 2 Aug. 2005 (Cleveland, OH)

Wine, Howard, 4 Sept. 2005 (Cleveland, OH)

Wingenfeld, Neal, 15 Sept. 2008 (Cleveland, OH)

· Website ·

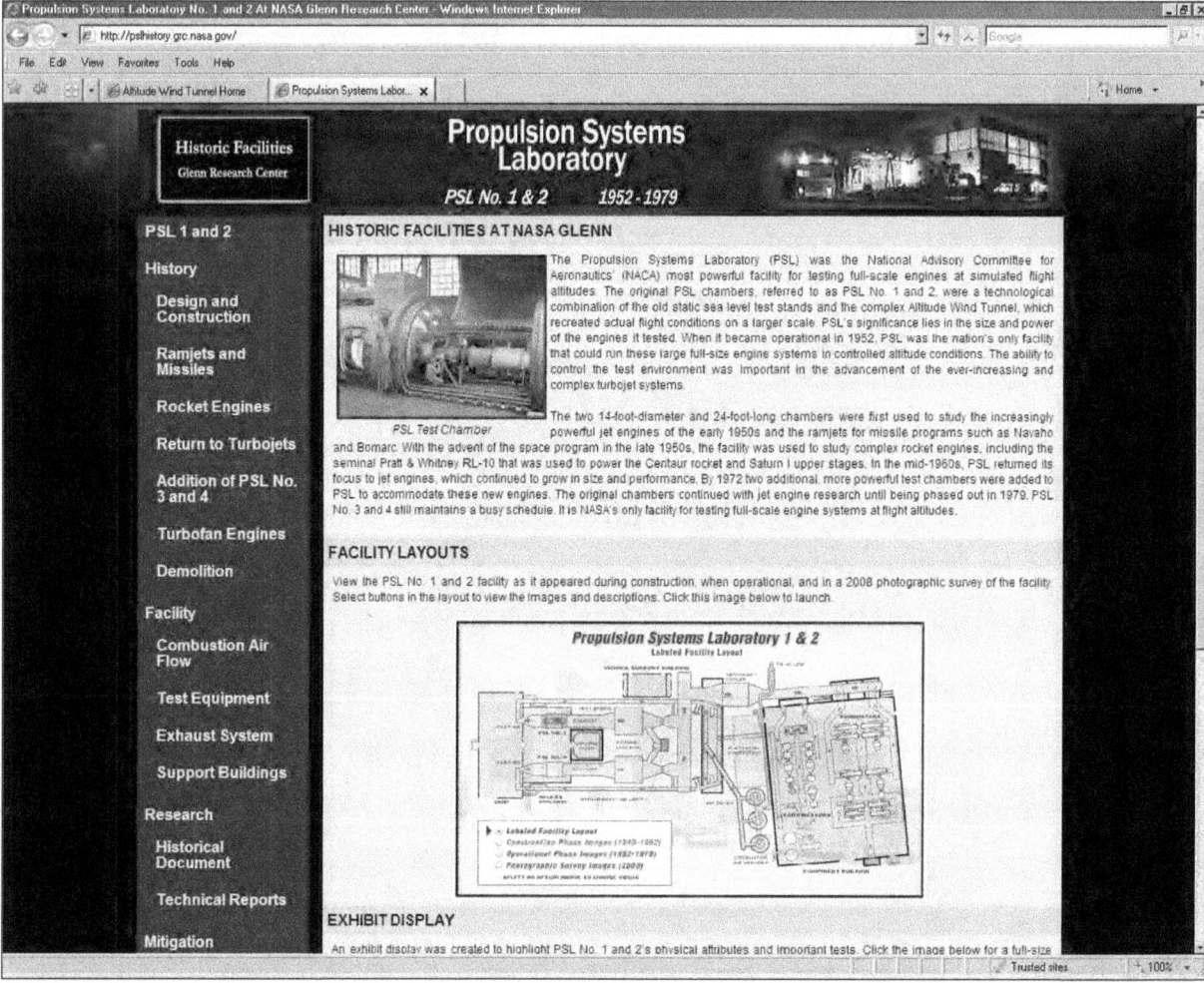

The documentation of the Propulsion Systems Laboratory (PSL) included the gathering of historical materials and information regarding the facility and its history. Many of these are available on the website *http://pslhistory.grc.nasa.gov*. The website includes an overview of PSL's history from its construction, study of ramjets and missiles, rocket engine testing, return to aeronautics, addition of PSL No. 3 and 4, and eventual demolition.

An interactive media piece highlights the facility's layout with photographs from the construction, operation, and final years. It comprises four sections:

The Facility section of the website describes the PSL's physical characteristics and operation in detail, including descriptions of the combustion air system, test equipment, control room, exhaust system, and support buildings. The Mitigation section illustrates the historical mitigation performed prior to demolition of the facility and includes related documents and photographs. The Research section contains almost 50 scanned documents and links to 20 related technical reports. The Students section includes a timeline of related events, timelines of testing at both PSL No. 1 and 2 and PSL No. 3 and 4, and a concise synopsis of how the PSL worked with a glossary of related terms.

· NASA History Series ·

Reference Works, NASA SP–4000 Series:

Grimwood, James M. *Project Mercury: A Chronology.* NASA SP–4001, 1963.

Grimwood, James M., and Barton C. Hacker, with Peter J. Vorzimmer. *Project Gemini Technology and Operations: A Chronology.* NASA SP–4002, 1969.

Link, Mae Mills. *Space Medicine in Project Mercury.* NASA SP–4003, 1965.

Astronautics and Aeronautics, 1963: Chronology of Science, Technology, and Policy. NASA SP–4004, 1964.

Astronautics and Aeronautics, 1964: Chronology of Science, Technology, and Policy. NASA SP–4005, 1965.

Astronautics and Aeronautics, 1965: Chronology of Science, Technology, and Policy. NASA SP–4006, 1966.

Astronautics and Aeronautics, 1966: Chronology of Science, Technology, and Policy. NASA SP–4007, 1967.

Astronautics and Aeronautics, 1967: Chronology of Science, Technology, and Policy. NASA SP–4008, 1968.

Ertel, Ivan D., and Mary Louise Morse. *The Apollo Spacecraft: A Chronology, Volume I, Through November 7, 1962.* NASA SP–4009, 1969.

Morse, Mary Louise, and Jean Kernahan Bays. *The Apollo Spacecraft: A Chronology, Volume II, November 8, 1962–September 30, 1964.* NASA SP–4009, 1973.

Brooks, Courtney G., and Ivan D. Ertel. *The Apollo Spacecraft: A Chronology, Volume III, October 1, 1964–January 20, 1966.* NASA SP–4009, 1973.

Ertel, Ivan D., and Roland W. Newkirk, with Courtney G. Brooks. *The Apollo Spacecraft: A Chronology, Volume IV, January 21, 1966–July 13, 1974.* NASA SP–4009, 1978.

Astronautics and Aeronautics, 1968: Chronology of Science, Technology, and Policy. NASA SP–4010, 1969.

Newkirk, Roland W., and Ivan D. Ertel, with Courtney G. Brooks. *Skylab: A Chronology.* NASA SP–4011, 1977.

Van Nimmen, Jane, and Leonard C. Bruno, with Robert L. Rosholt. *NASA Historical Data Book, Volume I: NASA Resources, 1958–1968.* NASA SP–4012, 1976; rep. ed. 1988.

Ezell, Linda Neuman. *NASA Historical Data Book, Volume II: Programs and Projects, 1958–1968.* NASA SP–4012, 1988.

Ezell, Linda Neuman. *NASA Historical Data Book, Volume III: Programs and Projects, 1969–1978.* NASA SP–4012, 1988.

Gawdiak, Ihor, with Helen Fedor. *NASA Historical Data Book, Volume IV: NASA Resources, 1969–1978.* NASA SP–4012, 1994.

Rumerman, Judy A. *NASA Historical Data Book, Volume V: NASA Launch Systems, Space Transportation, Human Spaceflight, and Space Science, 1979–1988.* NASA SP–4012, 1999.

Rumerman, Judy A. *NASA Historical Data Book, Volume VI: NASA Space Applications, Aeronautics and Space Research and Technology, Tracking and Data Acquisition/Support Operations, Commercial Programs, and Resources, 1979–1988.* NASA SP–4012, 1999.

Rumerman, Judy A. *NASA Historical Data Book, Volume VII: NASA Launch Systems, Space Transportation, Human Spaceflight, and Space Science, 1989–1998.* NASA SP–2009–4012, 2009.

There is no SP–4013.

Astronautics and Aeronautics, 1969: Chronology of Science, Technology, and Policy. NASA SP–4014, 1970.

Astronautics and Aeronautics, 1970: Chronology of Science, Technology, and Policy. NASA SP–4015, 1972.

Astronautics and Aeronautics, 1971: Chronology of Science, Technology, and Policy. NASA SP–4016, 1972.

Astronautics and Aeronautics, 1972: Chronology of Science, Technology, and Policy. NASA SP–4017, 1974.

Astronautics and Aeronautics, 1973: Chronology of Science, Technology, and Policy. NASA SP–4018, 1975.

Astronautics and Aeronautics, 1974: Chronology of Science, Technology, and Policy. NASA SP–4019, 1977.

Astronautics and Aeronautics, 1975: Chronology of Science, Technology, and Policy. NASA SP–4020, 1979.

Astronautics and Aeronautics, 1976: Chronology of Science, Technology, and Policy. NASA SP–4021, 1984.

Astronautics and Aeronautics, 1977: Chronology of Science, Technology, and Policy. NASA SP–4022, 1986.

Astronautics and Aeronautics, 1978: Chronology of Science, Technology, and Policy. NASA SP–4023, 1986.

Astronautics and Aeronautics, 1979–1984: Chronology of Science, Technology, and Policy. NASA SP–4024, 1988.

Astronautics and Aeronautics, 1985: Chronology of Science, Technology, and Policy. NASA SP–4025, 1990.

Noordung, Hermann. *The Problem of Space Travel: The Rocket Motor.* Edited by Ernst Stuhlinger and J. D. Hunley, with Jennifer Garland. NASA SP–4026, 1995.

Astronautics and Aeronautics, 1986–1990: A Chronology. NASA SP–4027, 1997.

Astronautics and Aeronautics, 1991–1995: A Chronology. NASA SP–2000–4028, 2000.

Orloff, Richard W. *Apollo by the Numbers: A Statistical Reference.* NASA SP–2000–4029, 2000.

Lewis, Marieke, and Ryan Swanson. *Astronautics and Aeronautics: A Chronology, 1996–2000.* NASA SP–2009–4030, 2009.

Ivey, William Noel, and Ryan Swanson. *Astronautics and Aeronautics: A Chronology, 2001–2005.* NASA SP–2010–4031, 2010.

Management Histories, NASA SP–4100 Series

Rosholt, Robert L. *An Administrative History of NASA, 1958–1963.* NASA SP–4101, 1966.

Levine, Arnold S. *Managing NASA in the Apollo Era.* NASA SP–4102, 1982.

Roland, Alex. *Model Research: The National Advisory Committee for Aeronautics, 1915–1958.* NASA SP–4103, 1985.

Fries, Sylvia D. *NASA Engineers and the Age of Apollo.* NASA SP–4104, 1992.

Glennan, T. Keith. *The Birth of NASA: The Diary of T. Keith Glennan.* Edited by J. D. Hunley. NASA SP–4105, 1993.

Seamans, Robert C. *Aiming at Targets: The Autobiography of Robert C. Seamans.* NASA SP–4106, 1996.

Garber, Stephen J., ed. *Looking Backward, Looking Forward: Forty Years of Human Spaceflight Symposium* NASA SP–2002–4107, 2002.

Mallick, Donald L., with Peter W. Merlin. *The Smell of Kerosene: A Test Pilot's Odyssey.* NASA SP–4108, 2003.

Iliff, Kenneth W., and Curtis L. Peebles. *From Runway to Orbit: Reflections of a NASA Engineer.* NASA SP–2004–4109, 2004.

Chertok, Boris. *Rockets and People, Volume I.* NASA SP–2005–4110, 2005.

Chertok, Boris. *Rockets and People: Creating a Rocket Industry, Volume II.* NASA SP–2006–4110, 2006.

Chertok, Boris. *Rockets and People: Hot Days of the Cold War, Volume III.* NASA SP–2009–4110, 2009.

Chertok, Boris. *Rockets and People: The Moon Race, Volume IV.* NASA SP–2011–4110, 2011.

Laufer, Alexander, Todd Post, and Edward Hoffman. *Shared Voyage: Learning and Unlearning from Remarkable Projects*. NASA SP–2005–4111, 2005.

Dawson, Virginia P., and Mark D. Bowles. *Realizing the Dream of Flight: Biographical Essays in Honor of the Centennial of Flight, 1903–2003*. NASA SP–2005–4112, 2005.

Mudgway, Douglas J. *William H. Pickering: America's Deep Space Pioneer*. NASA SP–2008–4113, 2008.

Project Histories, NASA SP–4200 Series

Swenson, Loyd S., Jr., James M. Grimwood, and Charles C. Alexander. *This New Ocean: A History of Project Mercury*. NASA SP–4201, 1966; rep. ed. 1999.

Green, Constance McLaughlin, and Milton Lomask. *Vanguard: A History*. NASA SP–4202, 1970; rep. ed. Smithsonian Institution Press, 1971.

Hacker, Barton C., and James M. Grimwood. *On the Shoulders of Titans: A History of Project Gemini*. NASA SP–4203, 1977; rep. ed. 2002.

Benson, Charles D., and William Barnaby Faherty. *Moonport: A History of Apollo Launch Facilities and Operations*. NASA SP–4204, 1978.

Brooks, Courtney G., James M. Grimwood, and Loyd S. Swenson, Jr. *Chariots for Apollo: A History of Manned Lunar Spacecraft*. NASA SP–4205, 1979.

Bilstein, Roger E. *Stages to Saturn: A Technological History of the Apollo/Saturn Launch Vehicles*. NASA SP–4206, 1980 and 1996.

There is no SP–4207.

Compton, W. David, and Charles D. Benson. *Living and Working in Space: A History of Skylab*. NASA SP–4208, 1983.

Ezell, Edward Clinton, and Linda Neuman Ezell. *The Partnership: A History of the Apollo-Soyuz Test Project*. NASA SP–4209, 1978.

Hall, R. Cargill. *Lunar Impact: A History of Project Ranger*. NASA SP–4210, 1977.

Newell, Homer E. *Beyond the Atmosphere: Early Years of Space Science*. NASA SP–4211, 1980.

Ezell, Edward Clinton, and Linda Neuman Ezell. *On Mars: Exploration of the Red Planet, 1958–1978*. NASA SP–4212, 1984.

Pitts, John A. *The Human Factor: Biomedicine in the Manned Space Program to 1980*. NASA SP–4213, 1985.

Compton, W. David. *Where No Man Has Gone Before: A History of Apollo Lunar Exploration Missions*. NASA SP–4214, 1989.

Naugle, John E. *First Among Equals: The Selection of NASA Space Science Experiments*. NASA SP–4215, 1991.

Wallace, Lane E. *Airborne Trailblazer: Two Decades with NASA Langley's 737 Flying Laboratory*. NASA SP–4216, 1994.

Butrica, Andrew J., ed. *Beyond the Ionosphere: Fifty Years of Satellite Communications*. NASA SP–4217, 1997.

Butrica, Andrew J. *To See the Unseen: A History of Planetary Radar Astronomy*. NASA SP–4218, 1996.

Mack, Pamela E., ed. *From Engineering Science to Big Science: The NACA and NASA Collier Trophy Research Project Winners*. NASA SP–4219, 1998.

Reed, R. Dale. *Wingless Flight: The Lifting Body Story*. NASA SP–4220, 1998.

Heppenheimer, T. A. *The Space Shuttle Decision: NASA's Search for a Reusable Space Vehicle*. NASA SP–4221, 1999.

Hunley, J. D., ed. *Toward Mach 2: The Douglas D–558 Program*. NASA SP–4222, 1999.

Swanson, Glen E., ed. *"Before This Decade Is Out . . ." Personal Reflections on the Apollo Program*. NASA SP–4223, 1999.

Tomayko, James E. *Computers Take Flight: A History of NASA's Pioneering Digital Fly-By-Wire Project*. NASA SP–4224, 2000.

Morgan, Clay. *Shuttle-Mir: The United States and Russia Share History's Highest Stage*. NASA SP–2001–4225, 2001.

Leary, William M. *"We Freeze to Please": A History of NASA's Icing Research Tunnel and the Quest for Safety*. NASA SP–2002–4226, 2002.

Mudgway, Douglas J. *Uplink-Downlink: A History of the Deep Space Network, 1957–1997*. NASA SP–2001–4227, 2001.

There is no SP–4228 or SP–4229.

Dawson, Virginia P., and Mark D. Bowles. *Taming Liquid Hydrogen: The Centaur Upper Stage Rocket, 1958–2002*. NASA SP–2004–4230, 2004.

Meltzer, Michael. *Mission to Jupiter: A History of the Galileo Project*. NASA SP–2007–4231, 2007.

Heppenheimer, T. A. *Facing the Heat Barrier: A History of Hypersonics*. NASA SP–2007–4232, 2007.

Tsiao, Sunny. *"Read You Loud and Clear!" The Story of NASA's Spaceflight Tracking and Data Network*. NASA SP–2007–4233, 2007.

Meltzer, Michael. *When Biospheres Collide: a History of NASA's Planetary Protection Programs*. NASA SP–2011–4234.

Center Histories, NASA SP–4300 Series

Rosenthal, Alfred. *Venture into Space: Early Years of Goddard Space Flight Center*. NASA SP–4301, 1985.

Hartman, Edwin P. *Adventures in Research: A History of Ames Research Center, 1940–1965*. NASA SP–4302, 1970.

Hallion, Richard P. *On the Frontier: Flight Research at Dryden, 1946–1981*. NASA SP–4303, 1984.

Muenger, Elizabeth A. *Searching the Horizon: A History of Ames Research Center, 1940–1976*. NASA SP–4304, 1985.

Hansen, James R. *Engineer in Charge: A History of the Langley Aeronautical Laboratory, 1917–1958*. NASA SP–4305, 1987.

Dawson, Virginia P. *Engines and Innovation: Lewis Laboratory and American Propulsion Technology*. NASA SP–4306, 1991.

Dethloff, Henry C. *"Suddenly Tomorrow Came . . .": A History of the Johnson Space Center, 1957–1990*. NASA SP–4307, 1993.

Hansen, James R. *Spaceflight Revolution: NASA Langley Research Center from Sputnik to Apollo*. NASA SP–4308, 1995.

Wallace, Lane E. *Flights of Discovery: An Illustrated History of the Dryden Flight Research Center*. NASA SP–4309, 1996.

Herring, Mack R. *Way Station to Space: A History of the John C. Stennis Space Center*. NASA SP–4310, 1997.

Wallace, Harold D., Jr. *Wallops Station and the Creation of an American Space Program*. NASA SP–4311, 1997.

Wallace, Lane E. *Dreams, Hopes, Realities. NASA's Goddard Space Flight Center: The First Forty Years.* NASA SP–4312, 1999.

Dunar, Andrew J., and Stephen P. Waring. *Power to Explore: A History of Marshall Space Flight Center, 1960–1990.* NASA SP–4313, 1999.

Bugos, Glenn E. *Atmosphere of Freedom: Sixty Years at the NASA Ames Research Center.* NASA SP–2000–4314, 2000.

There is no SP–4315.

Schultz, James. *Crafting Flight: Aircraft Pioneers and the Contributions of the Men and Women of NASA Langley Research Center.* NASA SP–2003–4316, 2003.

Bowles, Mark D. *Science in Flux: NASA's Nuclear Program at Plum Brook Station, 1955–2005.* NASA SP–2006–4317, 2006.

Wallace, Lane E. *Flights of Discovery: An Illustrated History of the Dryden Flight Research Center.* NASA SP–2007–4318, 2007. Revised version of NASA SP–4309.

Arrighi, Robert S. *Revolutionary Atmosphere: The Story of the Altitude Wind Tunnel and the Space Power Chambers.* NASA SP–2010–4319, 2010.

General Histories, NASA SP–4400 Series

Corliss, William R. *NASA Sounding Rockets, 1958–1968: A Historical Summary.* NASA SP–4401, 1971.

Wells, Helen T., Susan H. Whiteley, and Carrie Karegeannes. *Origins of NASA Names.* NASA SP–4402, 1976.

Anderson, Frank W., Jr. *Orders of Magnitude: A History of NACA and NASA, 1915–1980.* NASA SP–4403, 1981.

Sloop, John L. *Liquid Hydrogen as a Propulsion Fuel, 1945–1959.* NASA SP–4404, 1978.

Roland, Alex. *A Spacefaring People: Perspectives on Early Spaceflight.* NASA SP–4405, 1985.

Bilstein, Roger E. *Orders of Magnitude: A History of the NACA and NASA, 1915–1990.* NASA SP–4406, 1989.

Logsdon, John M., ed., with Linda J. Lear, Jannelle Warren Findley, Ray A. Williamson, and Dwayne A. Day. *Exploring the Unknown: Selected Documents in the History of the U.S. Civil Space Program, Volume I: Organizing for Exploration.* NASA SP–4407, 1995.

Logsdon, John M., ed., with Dwayne A. Day and Roger D. Launius. *Exploring the Unknown: Selected Documents in the History of the U.S. Civil Space Program, Volume II: External Relationships.* NASA SP–4407, 1996.

Logsdon, John M., ed., with Roger D. Launius, David H. Onkst, and Stephen J. Garber. *Exploring the Unknown: Selected Documents in the History of the U.S. Civil Space Program, Volume III: Using Space.* NASA SP–4407, 1998.

Logsdon, John M., ed., with Ray A. Williamson, Roger D. Launius, Russell J. Acker, Stephen J. Garber, and Jonathan L. Friedman. *Exploring the Unknown: Selected Documents in the History of the U.S. Civil Space Program, Volume IV: Accessing Space.* NASA SP–4407, 1999.

Logsdon, John M., ed., with Amy Paige Snyder, Roger D. Launius, Stephen J. Garber, and Regan Anne Newport. *Exploring the Unknown: Selected Documents in the History of the U.S. Civil Space Program, Volume V: Exploring the Cosmos.* NASA SP–2001–4407, 2001.

Logsdon, John M., ed., with Stephen J. Garber, Roger D. Launius, and Ray A. Williamson. *Exploring the Unknown: Selected Documents in the History of the U.S. Civil Space Program, Volume VI: Space and Earth Science.* NASA SP–2004–4407, 2004.

Logsdon, John M., ed., with Roger D. Launius. *Exploring the Unknown: Selected Documents in the History of the U.S. Civil Space Program, Volume VII: Human Spaceflight: Projects Mercury, Gemini, and Apollo.* NASA SP–2008–4407, 2008.

Siddiqi, Asif A., *Challenge to Apollo: The Soviet Union and the Space Race, 1945–1974.* NASA SP–2000–4408, 2000.

Hansen, James R., ed. *The Wind and Beyond: Journey into the History of Aerodynamics in America, Volume 1: The Ascent of the Airplane.* NASA SP–2003–4409, 2003.

Hansen, James R., ed. *The Wind and Beyond: Journey into the History of Aerodynamics in America, Volume 2: Reinventing the Airplane.* NASA SP–2007–4409, 2007.

Hogan, Thor. *Mars Wars: The Rise and Fall of the Space Exploration Initiative.* NASA SP–2007–4410, 2007.

Vakoch, Douglas A., editor. *Psychology of Space Exploration: Contemporary Research in Historical Perspective.* NASA SP–2011–4411, 2011.

Monographs in Aerospace History, SP–4500 Series

Launius, Roger D., and Aaron K. Gillette, comps. *Toward a History of the Space Shuttle: An Annotated Bibliography.* Monographs in Aerospace History, No. 1, 1992.

Launius, Roger D., and J. D. Hunley, comps. *An Annotated Bibliography of the Apollo Program.* Monographs in Aerospace History, No. 2, 1994.

Launius, Roger D. *Apollo: A Retrospective Analysis.* Monographs in Aerospace History, No. 3, 1994.

Hansen, James R. *Enchanted Rendezvous: John C. Houbolt and the Genesis of the Lunar–Orbit Rendezvous Concept.* Monographs in Aerospace History, No. 4, 1995.

Gorn, Michael H. *Hugh L. Dryden's Career in Aviation and Space.* Monographs in Aerospace History, No. 5, 1996.

Powers, Sheryll Goecke. *Women in Flight Research at NASA Dryden Flight Research Center from 1946 to 1995.* Monographs in Aerospace History, No. 6, 1997.

Portree, David S. F., and Robert C. Trevino. *Walking to Olympus: An EVA Chronology.* Monographs in Aerospace History, No. 7, 1997.

Logsdon, John M., moderator. *Legislative Origins of the National Aeronautics and Space Act of 1958: Proceedings of an Oral History Workshop.* Monographs in Aerospace History, No. 8, 1998.

Rumerman, Judy A., comp. *U.S. Human Spaceflight: A Record of Achievement, 1961–1998.* Monographs in Aerospace History, No. 9, 1998.

Portree, David S. F. *NASA's Origins and the Dawn of the Space Age.* Monographs in Aerospace History, No. 10, 1998.

Logsdon, John M. *Together in Orbit: The Origins of International Cooperation in the Space Station.* Monographs in Aerospace History, No. 11, 1998.

Phillips, W. Hewitt. *Journey in Aeronautical Research: A Career at NASA Langley Research Center.* Monographs in Aerospace History, No. 12, 1998.

Braslow, Albert L. *A History of Suction-Type Laminar-Flow Control with Emphasis on Flight Research.* Monographs in Aerospace History, No. 13, 1999.

Logsdon, John M., moderator. *Managing the Moon Program: Lessons Learned from Apollo.* Monographs in Aerospace History, No. 14, 1999.

Perminov, V. G. *The Difficult Road to Mars: A Brief History of Mars Exploration in the Soviet Union*. Monographs in Aerospace History, No. 15, 1999.

Tucker, Tom. *Touchdown: The Development of Propulsion Controlled Aircraft at NASA Dryden*. Monographs in Aerospace History, No. 16, 1999.

Maisel, Martin, Demo J. Giulanetti, and Daniel C. Dugan. *The History of the XV-15 Tilt Rotor Research Aircraft: From Concept to Flight*. Monographs in Aerospace History, No. 17, 2000. NASA SP–2000–4517.

Jenkins, Dennis R. *Hypersonics Before the Shuttle: A Concise History of the X-15 Research Airplane*. Monographs in Aerospace History, No. 18, 2000. NASA SP–2000–4518.

Chambers, Joseph R. *Partners in Freedom: Contributions of the Langley Research Center to U.S. Military Aircraft of the 1990s*. Monographs in Aerospace History, No. 19, 2000. NASA SP–2000–4519.

Waltman, Gene L. *Black Magic and Gremlins: Analog Flight Simulations at NASA's Flight Research Center*. Monographs in Aerospace History, No. 20, 2000. NASA SP–2000–4520.

Portree, David S. F. *Humans to Mars: Fifty Years of Mission Planning, 1950–2000*. Monographs in Aerospace History, No. 21, 2001. NASA SP–2001–4521.

Thompson, Milton O., with J. D. Hunley. *Flight Research: Problems Encountered and What They Should Teach Us*. Monographs in Aerospace History, No. 22, 2001. NASA SP–2001–4522.

Tucker, Tom. *The Eclipse Project*. Monographs in Aerospace History, No. 23, 2001. NASA SP–2001–4523.

Siddiqi, Asif A. *Deep Space Chronicle: A Chronology of Deep Space and Planetary Probes, 1958–2000*. Monographs in Aerospace History, No. 24, 2002. NASA SP–2002–4524.

Merlin, Peter W. *Mach 3+: NASA/USAF YF-12 Flight Research, 1969–1979*. Monographs in Aerospace History, No. 25, 2001. NASA SP–2001–4525.

Anderson, Seth B. *Memoirs of an Aeronautical Engineer: Flight Tests at Ames Research Center: 1940–1970*. Monographs in Aerospace History, No. 26, 2002. NASA SP–2002–4526.

Renstrom, Arthur G. *Wilbur and Orville Wright: A Bibliography Commemorating the One-Hundredth Anniversary of the First Powered Flight on December 17, 1903*. Monographs in Aerospace History, No. 27, 2002. NASA SP–2002–4527.

There is no monograph 28.

Chambers, Joseph R. *Concept to Reality: Contributions of the NASA Langley Research Center to U.S. Civil Aircraft of the 1990s*. Monographs in Aerospace History, No. 29, 2003. NASA SP–2003–4529.

Peebles, Curtis, ed. *The Spoken Word: Recollections of Dryden History, The Early Years*. Monographs in Aerospace History, No. 30, 2003. NASA SP–2003–4530.

Jenkins, Dennis R., Tony Landis, and Jay Miller. *American X-Vehicles: An Inventory—X-1 to X-50*. Monographs in Aerospace History, No. 31, 2003. NASA SP–2003–4531.

Renstrom, Arthur G. *Wilbur and Orville Wright: A Chronology Commemorating the One-Hundredth Anniversary of the First Powered Flight on December 17, 1903*. Monographs in Aerospace History, No. 32, 2003. NASA SP–2003–4532.

Bowles, Mark D., and Robert S. Arrighi. *NASA's Nuclear Frontier: The Plum Brook Research Reactor*. Monographs in Aerospace History, No. 33, 2004. NASA SP–2004–4533.

Wallace, Lane, and Christian Gelzer. *Nose Up: High Angle-of-Attack and Thrust Vectoring Research at NASA Dryden, 1979–2001*. Monographs in Aerospace History, No. 34, 2009. NASA SP–2009–4534.

Matranga, Gene J., C. Wayne Ottinger, Calvin R. Jarvis, and D. Christian Gelzer. *Unconventional, Contrary, and Ugly: The Lunar Landing Research Vehicle*. Monographs in Aerospace History, No. 35, 2006. NASA SP–2004–4535.

McCurdy, Howard E. *Low-Cost Innovation in Spaceflight: The History of the Near Earth Asteroid Rendezvous (NEAR) Mission*. Monographs in Aerospace History, No. 36, 2005. NASA SP–2005–4536.

Seamans, Robert C., Jr. *Project Apollo: The Tough Decisions*. Monographs in Aerospace History, No. 37, 2005. NASA SP–2005–4537.

Lambright, W. Henry. *NASA and the Environment: The Case of Ozone Depletion*. Monographs in Aerospace History, No. 38, 2005. NASA SP–2005–4538.

Chambers, Joseph R. *Innovation in Flight: Research of the NASA Langley Research Center on Revolutionary Advanced Concepts for Aeronautics*. Monographs in Aerospace History, No. 39, 2005. NASA SP–2005–4539.

Phillips, W. Hewitt. *Journey into Space Research: Continuation of a Career at NASA Langley Research Center*. Monographs in Aerospace History, No. 40, 2005. NASA SP–2005–4540.

Rumerman, Judy A., Chris Gamble, and Gabriel Okolski, comps. *U.S. Human Spaceflight: A Record of Achievement, 1961–2006*. Monographs in Aerospace History, No. 41, 2007. NASA SP–2007–4541.

Peebles, Curtis. *The Spoken Word: Recollections of Dryden History Beyond the Sky*. Monographs in Aerospace History, No. 42, 2011. NASA SP–2011–4542.

Dick, Steven J., Stephen J. Garber, and Jane H. Odom. *Research in NASA History*. Monographs in Aerospace History, No. 43, 2009. NASA SP–2009–4543.

Merlin, Peter W. *Ikhana: Unmanned Aircraft System Western States Fire Missions*. Monographs in Aerospace History, No. 44, 2009. NASA SP–2009–4544.

Fisher, Steven C., and Shamim A. Rahman. *Remembering the Giants: Apollo Rocket Propulsion Development*. Monographs in Aerospace History, No. 45, 2009. NASA SP–2009–4545.

Gelzer, Christian. *Fairing Well: From Shoebox to Bat Truck and Beyond, Aerodynamic Truck Research at NASA's Dryden Flight Research Center*. Monographs in Aerospace History, No. 46, 2011. NASA SP–2011–4546.

Electronic Media, SP–4600 Series

Remembering Apollo 11: The 30th Anniversary Data Archive CD–ROM. NASA SP–4601, 1999.

Remembering Apollo 11: The 35th Anniversary Data Archive CD–ROM. NASA SP–2004–4601, 2004. This is an update of the 1999 edition.

The Mission Transcript Collection: U.S. Human Spaceflight Missions from Mercury Redstone 3 to Apollo 17. NASA SP–2000–4602, 2001.

Shuttle-Mir: The United States and Russia Share History's Highest Stage. NASA SP–2001–4603, 2002.

U.S. Centennial of Flight Commission Presents Born of Dreams—Inspired by Freedom. NASA SP–2004–4604, 2004.

Of Ashes and Atoms: A Documentary on the NASA Plum Brook Reactor Facility. NASA SP–2005–4605, 2005.

Taming Liquid Hydrogen: The Centaur Upper Stage Rocket Interactive CD-ROM. NASA SP–2004–4606, 2004.

Fueling Space Exploration: The History of NASA's Rocket Engine Test Facility DVD. NASA SP–2005–4607, 2005.

Altitude Wind Tunnel at NASA Glenn Research Center: An Interactive History CD-ROM. NASA SP–2008–4608, 2008.

A Tunnel Through Time: The History of NASA's Altitude Wind Tunnel. NASA SP–2010–4609, 2010.

Conference Proceedings, SP–4700 Series

Dick, Steven J., and Keith Cowing, eds. *Risk and Exploration: Earth, Sea and the Stars.* NASA SP–2005–4701, 2005.

Dick, Steven J., and Roger D. Launius. *Critical Issues in the History of Spaceflight.* NASA SP–2006–4702, 2006.

Dick, Steven J., ed. *Remembering the Space Age: Proceedings of the 50th Anniversary Conference.* NASA SP–2008–4703, 2008.

Dick, Steven J., ed. *NASA's First 50 Years: Historical Perspectives.* NASA SP–2010–4704, 2010.

Societal Impact, SP–4800 Series

Dick, Steven J., and Roger D. Launius. *Societal Impact of Spaceflight.* NASA SP–2007–4801, 2007.

Dick, Steven J., and Mark L. Lupisella. *Cosmos and Culture: Cultural Evolution in a Cosmic Context.* NASA SP–2009–4802, 2009.

· List of Images ·

Cover: A NASA technician examines a propellant tank for the Weightless Analysis Sounding Probe (WASP) setup inside the PSL No. 2 test chamber. (NASA C–1962–60067)

Introduction
1. PSL No. 1 and 2 exterior at night (NASA C–1952–30764), p. vii.
2. PSL aerial from the west (NASA C–1959–50861), p. viii.
3. Exhibit display describing PSL's history (NASA P0931 PSL 1 2), p. x

Chapter 1
4. Test of an engine in the Prop House (NASA C–1944–04598), p. 2.
5. Pilot Shorty Schroeder (U.S. Air Force Museum), p. 3.
6. William Stratton (NASA A–6786), p. 4
7. National Bureau of Standards campus (National Institute of Standards and Technology), p. 4.
8. NBS Altitude Laboratory (National Institute of Standards and Technology), p. 5.
9. Dynamometer in Engine Research Building (NASA C–1943–01345), p. 7.
10. Ramjet in Altitude Wind Tunnel (NASA C–1946–14244), p. 8.
11. Four Burner Area in Engine Research Building (NASA C–1947–19780), p. 9.
12. General Electric J73 engine in PSL No. 1 (NASA C–1952–30959), p. 10.

Chapter 2
13. PSL construction (NASA C–2009–00180), p. 12.
14. Abe Silverstein and Ray Sharp (NASA C–1953–34030), p. 13.
15. Abe Silverstein, Oscar Schey, Benjamin Pinkel, Jesse Hall, and John Collins (NASA C–1949–22821), p. 14.
16. Model of PSL No. 1 and 2 (NASA C–1954–37139), p. 15.
17. Eugene Wasielewski (NASA C–1956–42127), p. 17.
18. Construction of cooler stands (NASA C–2009–00175), p. 19.
19. Assembling of overhead air pipe (NASA C–2009–00181), p. 20.
20. Aerial view of site before PSL construction (NASA C–1949–24972), p. 20.
21. Support for exhaust cooler (NASA C–2009–00151), p. 21.
22. Excavations for the Equipment Building (NASA C–2009–00183), p. 21.
23. Building of an exhaust gas cooler (NASA C–2009–00177), p. 22.
24. Construction of Shop and Access Building (NASA C–2009–00179), p. 23.
25. Long view of construction site (NASA C–2009–00184), p. 23.
26. PSL altitude chambers arrive on trucks (NASA C–2009–00173), p. 24.
27. Chamber inlet section hoisted on crane (NASA C–2009–00171), p. 24.
28. Construction of Shop and Access Building (NASA C–2009–00195), p. 25.
29. Construction of Equipment Building (NASA C–2009–00166), p. 25.
30. Elliott compressors being installed (NASA C–1952–29500), p. 26.
31. Construction of air heaters (NASA C–2009–00163), p. 27.
32. Wooden framework for the cooling tower (NASA C–2009–00161), p. 27.
33. Bud Meilander (NASA C–1966–01760), p. 28.
34. Lyle Dickerhoff (NASA C–1958–48725), p. 28.
35. Carl Betz (NASA C–1956–42477), p. 28.
36. PSL exterior at night (NASA C–1952–30767), p. 29.

Pursuit of Power

Chapter 3

37. Model of PSL No. 1 and 2 (NASA C–1954–37141), p. 32.
38. Seymour Himmel at 1954 Inspection (NASA C–1954–36018), p. 33.
39. Floor plan for PSL 1954 Inspection talks (NASA), p. 34.
40. Three-dimensional drawing of PSL No. 1 and 2 (NASA CD–09–83168), p. 35.
41. Martin Saari and another research engineer (NASA C–1957–45797), p. 37.
42. Test Installations Division staff with engine (NASA C–1967–01582), p. 37.
43. Engine beside PSL test chamber (NASA C–1955–40070), p. 38.
44. Test engineer and mechanic (NASA C–1957–45798), p. 38.
45. Direct-connect setup in chamber (NASA C–1955–40235), p. 39.
46. Free-jet setup in chamber (NASA C–1954–36894), p. 39.
47. Mechanic installing instrumentation on engine (NASA C–1977–03128), p. 39.
48. Cutaway drawing of PSL Shop and Access Building (NASA 8090EL), p. 40.
49. First shift engineers in PSL chamber (NASA C–1953–33436), p. 40.
50. Third shift operators outside the PSL (NASA C–1952–30766), p. 41.
51. Exterior of the PSL at night (NASA C–1952–30768), p. 41.
52. PSL No. 1 with lid closed (NASA C–2008–04129), p. 42.
53. Equipment Building control room (NASA C–1953–31863), p. 42.
54. Elliott compressors in Equipment Building (NASA C–1954–35700), p. 43.
55. Exterior of PSL Shop and Access Building (NASA C–1953–31941), p. 43.
56. Air heaters outside Equipment Building (NASA C–1952–30936), p. 44.
57. Refrigeration equipment inside Equipment Building (NASA C–1953–3228), p. 44.
58. Diagram of PSL test chamber (NACA RM E53I08, fig. 3), p. 45.
59. PSL No. 2 with doors open (NASA C–1952–30962), p. 45.
60. Howard Wine and operators in the control room (NASA C–1957–45647), p. 46.
61. Wide view of control room with Bob Walker (NASA C–1954–35900), p. 46.
62. Control room in 1973 (NASA C–1973–01519), p. 47.
63. Control room entrance (NASA C–1963–63291), p. 47.
64. Exhausters inside Equipment Building (NASA C–1959–49428), p. 48.
65. Technician examines exhauster (NASA C–1963–65822), p. 48.
66. Exterior of a primary cooler (NASA C–1952–30771), p. 49.
67. Fins inside a primary cooler (NASA C–1959–49583), p. 49.
68. Exterior of secondary cooler (NASA C–1952–30937), p. 50.
69. Cooling tower (NASA C–1952–30938), p. 50.

Chapter 4

70. Researcher with a J79 turbojet (NASA C–1956–42401), p. 52.
71. Carlton Kemper (NASA C–1943–01297), p. 53.
72. Drawing of Navaho missile in flight (NASA C–1956–43440), p. 54.
73. Navaho X-10 in flight (NASA M–9142273), p. 55.
74. Bob Walker (NASA C–1953–34978), p 56.
75. Damaged combustion liner (NASA C–1955–37979), p. 57.
76. View through chamber showing mechanics inside (NASA C–1952–30950), p. 58.
77. Dwight Reilly examining XRJ47 setup (NASA C–1953–33390), p. 59.
78. Collage of different flameholder designs (NASA C–1954–35354, C–1952–31065, C–1954–36486, C–1952–31248, C–1955–39035, C–1952–31066, C–1954–36745, C–1952–31064, C–1955–40072, C–1954–35888, C–1956–41467, and C–1954–35889), p. 60.
79. Bob Walker examines free-jet setup for XRJ47 tests (NASA C–1953–33387), p. 61.
80. 48-inch-diameter ramjet in PSL No. 2 (NASA C–1952–30961), p. 62.

81. Navaho missile on display (Bob Walker), p. 63.
82. Gene Wasielewski (NASA C–1955–39823), p. 63.
83. Squadron of Bomarc missiles (U.S. Air Force), p. 64.
84. 28-inch-diameter ramjet in PSL No. 1 (NASA C–1954–36898), p. 65.
85. Howard Wine (NASA C–1965–01796), p. 66.
86. Martin Saari with J79 Inspection display (NASA C–1957–46136), p. 67.
87. J73 compressor blades (NASA C–1953–34173), p. 68.
88. F-4 Phantom in flight (U.S. Air Force), p. 68.
89. Avro Arrow fighter brake testing (Canadian Department of National Defense), p. 69.
90. John McAulay and technician install the Iroquois in the PSL (NASA C–1957–45206), p. 70.

Chapter 5

91. John Kobak and engineer examine drawing next to test chamber (NASA C–1962–62465), p. 74.
92. The PSL with surrounding buildings (NASA C–1955–38660), p. 76.
93. Technician with rocket in test chamber (NASA C–1958–48831), p. 77.
94. Diagram of pebble bed heater (NASA), p. 78.
95. Pebble bed heater installation in the PSL (NASA C–1958–49138), p. 79.
96. John Kobak (NASA C–1962–64266), p. 80.
97. Neal Wingenfeld (NASA C–1982–01508), p. 81.
98. Rocket component display (NASA C–1954–35975), p. 82.
99. John Kobak with wrecked rocket engine (NASA C–1960–53702), p. 83.
100. PSL control room for RL-10 tests (NASA C–1964–69528), p. 83.
101. Atlas/Centaur on the launch pad (NASA C–1964–68973), p. 84.
102. Ali Mansour and Ned Hannum pose with RL-10 engine (NASA C–1963–64329), p. 85.
103. Diagram of Centaur structure (NASA C–1953–34978), p. 85.
104. John Kobak and a technician examine RL-10 setup (NASA C–1962–60071), p. 86.
105. John Kobak and Gene Baughman examine RL-10 in PSL No. 1 (NASA C–1961–55596), p. 87.
106. Propellant tanks outside the PSL facility (NASA C–1963–64268), p. 89.
107. Engineer in PSL control room (NASA C–1963–67414), p. 89.
108. Large area ratio nozzles on display in the PSL (NASA C–1964–69633), p. 90.
109. Wayne Thomas (NASA C–1958–49066), p. 91.
110. Apollo contour nozzle inside PSL No. 1 (NASA C–1964–68789), p. 92.
111. 260-inch-diameter rocket engine assembly (NASA C–1965–03064), p. 93.
112. Bill Niedergard inspects solid rocket engine in PSL No. 2 (NASA C–1965–1037), p. 93.
113. Inspection display inside the PSL (NASA C–1966–03902), p. 94.

Chapter 6

114. Stator vanes in TF30 turbofan engine (NASA C–1967–02827), p. 98.
115. TF30 turbofan next to the PSL test chamber (NASA C–1967–03560), p. 100.
116. Engineers in the PSL (NASA C–1967–01584), p. 101.
117. Frank Kutina (NASA C–1957–44592), p. 101.
118. Bud Meilander examines flamespreader in PSL No. 2 (NASA C–1967–01733), p. 102.
119. F-106B Delta Dart (NASA C–1969–02871), p. 103.
120. Lee Wagoner examines J85-13 engine setup in PSL No. 2 (NASA C–1967–02128), p. 104.
121. Pratt & Whitney TF30 turbofan (NASA C–1967–02809), p. 105.
122. Leon Wenzel (NASA C–1957–45544), p. 106.
123. Robert Cummings and Harold Gold with Low Cost Engine (NASA C–1972–00577), p. 107.
124. Compass Cope Tern drone aircraft, p. 108.
125. ATF3 engine in the PSL (NASA C–1973–00344), p. 109.

126. Exterior of Equipment Building showing explosion damage (NASA C–1971–01269), p. 110.
127. Interior of Equipment Building showing explosion damage (NASA C–1971–01272), p. 111.
128. Blown header inside Equipment Building (NASA C–1971–01283), p. 112.

Chapter 7

129. Installation of test chamber in the new PSL facility (NASA C–1969–02672), p. 116.
130. Aerial of PSL site in 1945 (C–1945–13061), p. 118.
131. Aerial of PSL site in 1957 (NASA C–1957–44883), p. 118.
132. Aerial of PSL site in 1987 (C–1987–09393), p. 118.
133. Bulldozing of Wright Park for the new PSL facility (NASA C–1967–02211), p. 119.
134. Drawing of PSL No. 3 and 4 (NASA Drawing), p. 119.
135. Construction of primary cooler for PSL No. 3 and 4 (NASA C–1969–03898), p. 120.
136. Valve for new primary cooler for PSL No. 3 and 4 (NASA C–1970–01642), p. 120.
137. Overall view of construction of new test chambers (NASA C–1970–01312), p. 121.
138. Aerial of PSL No. 3 and 4 (NASA C–2006–01485), p. 121.
139. Updated PSL control room (NASA C–1970–01743), p. 122.
140. New line of exhausters (NASA C–1974–01139), p. 122.
141. Drawing of entire PSL complex (NASA CD–11–83305), p. 123.
142. Aerial view of entire PSL complex (NASA C–1974–03577), p. 124.
143. F-16 Falcon in flight (2003, U.S. Air Force), p. 125.
144. Technicians work on F100 engine in the PSL (NASA C–1978–00476), p. 126.
145. J85-21 in PSL No. 2 (NASA C–1974–01012), p. 127.
146. HiMAT aircraft (NASA C–1980–03906), p. 128.
147. Dryden pilot Gary Krier (NASA ECN–3091), p. 129.
148. PSL validation test chart (NASA C–1983–02762), p. 130.
149. Dryden's F-15A aircraft (NASA ECN–18899), p. 131.
150. PSL No. 1 and 2 test schedule for 1972–74 (NASA), p. 132.

Chapter 8

151. Primary cooler being demolished (NASA C–2009–02016), p. 136.
152. Overall view of interior of Shop and Access Building (NASA C–2008–04155), p. 138.
153. Atmospheric air vent in decay (NASA C–2008–04148), p. 139.
154. Cannibalized PSL control room (NASA C–2008–04125), p. 140.
155. Rubble at former PSL site (NASA C–2009–02018), p. 141.
156. PSL as it appeared in 1952 (NASA C–1952–30764), p. 141.
157. PSL piping prior to demolition (NASA C–2008–04472), p. 142.
158. Service Support Building (NASA C–2008–04132), p. 142.
159. Demolition of Shop and Access Building (NASA C–2009–01660), p. 143.
160. Demolition of primary cooler (NASA C–2009–02021), p. 143.
161. Demolition of PSL test chamber (NASA C–2009–01688), p. 144.
162. Demolition of PSL test chamber (NASA C–2009–01692), p. 144.
163. Joe Morris and Lisa Adkins tour PSL control room (NASA C–2008–04153), p. 145.
164. Howard Wine inside a PSL test chamber (NASA C–1962–60068), p. 146.
165. Interactive media piece showing PSL layout (NASA, available at http://pslhistory.grc.nasa.gov), p. 147.

Index

A

Access Building. *See* Propulsion Systems Laboratory, support buildings, Shop and Access Building
Ad Hoc Aeronautics Assessment Committee 126
Adkins, Lisa **145**
Advisory Council on Historic Preservation. *See* Historical preservation
Aeroelasticity of Turbine Engines 125
Aerojet 64, 75, 92, 94
    M-1 engine 80
    Test facilities 80, 92, 94
    260-inch rocket 92, **93**, 94
Afterburner. *See* Engine components (aircraft)
Airbreathing engines. *See* Engines; Engine components (aircraft); Engine problems (aircraft)
Aircraft
    Commercial 68, 70
    General aviation 107, 108
    Hypersonic 76, 77
    Supersonic 68, 77, 104. *See also* Convair, B-58 Hustler
    Vertical and/or short takeoff and landing 130
    See also Avro Aircraft Limited; Boeing; Grumman; Lockheed; McDonnell Douglas; North American Aviation; Northrup; Rockwell International; Teledyne Ryan Technologies Inc.; Vought
Aircraft Engine Research Laboratory (AERL). *See* Glenn Research Center, former names
Aircraft engines. *See* Engines; Engine components (aircraft); Engine problems (aircraft)
Air Force, U.S. 63, 65, 68, 75, 92, 105, 109, 110, 125, 126, 128, 129, 130, 131, 132
    Aircraft. *See* General Dynamics
    Edwards Air Force Base **108**
Aldershof. *See* Germany, test facilities
Allison
    V-1710 7
Alsos Mission (battle for German technology)
    Operation LUSTY (documents and hardware) 53
    Operation Paperclip (scientists) 53
Altitude research
    Pilots 3
    Problems. *See* Engine problems (aircraft), altitude; flameout (blow out); ignition
    Records 3
    Test chambers viii, ix, 3, 4, **5**, 6, 8, 9
Altitude Wind Tunnel (AWT). *See* Glenn Research Center, facilities
Ames Research Center (Ames Aeronautical Laboratory) 6, 127, 128
Apollo digital control system 129
Apollo Program 75, 77, 85, 99, 129, 137
Apollo Space Propulsion System **90**, 91
Army, U.S. 3, 85
Army Ballistic Missile Agency. *See* Marshall Space Flight Center
Arnold Engineering Development Center (AEDC) 69, 99
Atlas (rocket) 63, **74**, **84**, 85

Atomic weapons. *See* Alsos Mission
Auckerman, Carl 91
Avro Aircraft Limited 69, 70
    Arrow fighter 68, **69**
    Orenda Branch 68
    PS.13 Iroquois engine 68, 69, **70**, 76
AWT. *See* Glenn Research Center, facilities, Altitude Wind Tunnel

B

Baldwin-Wallace College 66
Baughman, Eugene 80, **87**
Betz, Carl **28**, 66
Bloomer, Harry 91
Bobula, George 128
Boeing 64, 65, 103, 104, 129
    B-47 69
    B-52 127
    Bomarc (missile) **64**, 65
    Supersonic Transport Technology (SST) 99, 103, 104
    YQM-94 B-Gulls 108
Braig, James 16, 28
Budget and Accounting Procedures Act of 1950 13, 15
Burkardt, Leo 128
Burns and Roe, Inc. 16

C

Calmer, Edward 32
Canadian aerospace engineers. *See* Avro Aircraft Limited
Cape Kennedy **84**
Case Institute of Technology 80, 81
Caswell, Bridget 137
Centaur (upper-stage rocket) 74, 75, **84, 85**, 87, 88, 94
    Centaur missions 85, 91
    Centaur Program 80, 91
    Failures 88
Central Air and Equipment Building. *See* Propulsion Systems Laboratory, support buildings, Equipment Building
Chemistry Laboratory. *See* Glenn Research Center, facilities
Childs, J. Howard 100
Chrysler 80
Cleveland vii, 7, 75
Cleveland Laboratory. *See* Glenn Research Center
Cleveland Municipal Airport 117
Cold War vii, 29, 53-54, 64
Collier Trophy 68
Collier's Magazine vii
Collins, John **14**
Collopy, Paul 132
Compass Cope. *See* Teledyne Ryan Technologies Inc., YQM-98 A Compass Cope Tern

Computer simulations 125, **130**
Congress 4, 6, 15, 65, 104
Conrad, William 33
Convair
    B-58 Hustler 52, 68
    F-106B Delta Dart **103**, 104
Crowl, Robert 65
Cryogenic. *See* Fuels, high energy, liquid hydrogen; liquid nitrogen
Cummings, Robert **107**
Curtiss-Wright 17, 54, 64. *See also* Wright Aeronautical Division

D
Dassault
    Falcon 110
Dawson, Virginia 7, 16, 149
Department of Energy. 80
Department of Transportation 99
Dickerhoff, Lyle **28**
Dickinson, Hobart Cutler 5, 6
Documentation. *See* Historical preservation
Drag. *See* Engine problems (aircraft)
Drone aircraft 107, 108. *See also* Teledyne Ryan Technologies Inc., YQM-98 A Compass Cope Tern; Rockwell International, HiMAT
Dryden Flight Research Center 127, 128, 129, 130, 131, 132
    Muroc Lake 15
Dunbar, William 62, 65
Dynamotor Laboratory. *See* National Bureau of Standards

E
Ebert, William 53
Egan, William 28, 66
Eisenhower, Dwight vii
Elliott Company 16, 26, 43
Employees. *See* Glenn Research Center, staff; Propulsion Systems Laboratory, staff
Engine components (aircraft) 53
    Afterburner 68, 69, 106, 129, 131
    Axial-flow compressor. *See* Turbojet engines, axial flow
    Combustor 61, 67
        Diffuser 62, 91
    Communications (digital) 131
    Compressor 67, 69, 105, 106, 124, 125, 129
        Compressor casing "treatments" 106
    Controls 128, 129. *See also* Pratt & Whitney
        Digital controls 127, 129, 130
            Digital Electronic Engine Control system 131
            Full Authority Digital Engine Control (FADEC) 131, 132
        Electronic Engine Control 110, 124, 129
        Fly-by-wire 127, 128-129
        Hydromechanical control 127, 129

Engine components (aircraft) cont.
　Controls cont.
　　Integrated Propulsion Control System (IPCS) 129, 130
　　Linear quadratic regulator 130, 131
　　Variable-geometry controls 129
　　　Multivariable control systems (MVCS or "smart carburetor") 129, 130, 131
　Cooling liner (burner shell) **57**
　Ejector assembly 68, 104, 110
　Flameholder 54-55, 57, **58**, **60**, 68
　　Types of 58, **60**, 61
　Fuel-distribution system 61, 68, 69, 110
　Igniter 57, 92
　Inlets 103, 104, 105, 110, 128, 129
　　Resonators 88
　Nozzles 103, 104, 110, 129
　　Supersonic. *See* General Electric
　Stator blades **68**, 124, 129
　Turbine 67
　Turbosupercharger 3, 5, 6, 17
Engine components (rocket) **82**
　Combustion chamber 91, 92
　Ejector assembly 68, 91
　Hydrostat 93
　Injectors 91, 92, **94**
　Nozzles 67, 68, 69, 92
　　Contour 91, **92**
　　Gimballing 92
　　　Aft thrust vector control 92
　　　Thrust vectoring 78, 92
　　Isentropic 77
　　Large-area-ratio 82, **90**, 91, 92
　　Low-area-ratio 91, 92
　Pressure chamber 92
　Turbopump 88
Engine problems (aircraft)
　Altitude viii, 3, 5, 6, 8, 108
　Component overheating 67
　Compressor aerodynamics 126
　Compressor efficiency 68, 107
　Compressor stability 105, 106, 125, 127
　Compressor stall 69, 105, 106, 110, 124, 126, 127
　Controls viii, 67, 99
　　Indicial and frequency response 62
　Cost 107. *See also* Low Cost Engine
　Distortion (distorted airflow) viii, 58, 67, 69, 100, 105, 106, 124, 126, 128. *See also* Pratt & Whitney
　Drag 103
　Emissions 99, 107
　Flameout (blow out) 8, 67
　Flutter viii, 99, 124, 125, 126, 127. *See also* Pratt & Whitney; General Electric

Engine problems (aircraft) cont.
    Fuel consumption and control 6, 61, 99, 103, 104, 107, 110, 129. *See also* Fuels
    High Mach numbers 63
    Hot gases 34, 88, 92
    Ignition, 57, 58, 63, 69
    Low Reynolds numbers 105, 106
    Noise viii, 99, 103, 104, 110
    Pressure drop 68
    Thrust 104, 128
    Vacuum tube failure 129
    Windmilling 69
Engine problems (rocket)
    Ablative thrust chamber durability 91
    Annular channel flow 92
    Chugging 88
    Combustion instability 88, 91, 92
    Fuel consumption and control 88. *See also* Fuels
    Gimballing (steering) 87, 92
    Nozzle efficiency 91, 92
    Oscillations at low thrust 88
    Prechilling 88
    Screech 88
    Throttling 88
    Thrust 94
    Turbopumps, 88
Engine Propeller Research Building. *See* Glenn Research Center, facilities
Engine Research Building. *See* Glenn Research Center, facilities
Engines. *See* Allison; Garrett Corporation; General Dynamics; General Electric; Jet engines; Low Cost Engine; Marquardt; Pratt & Whitney; Quiet Engine; Ramjets; Reciprocating engines; Rockets, chemical rockets, solid rockets; Stirling engine; Turbofan engines; Turbojet engines; Westinghouse; Wright Aeronautical Division
ERB. *See* Glenn Research Center, facilities, Engine Research Building
European aerospace research 4. *See also* France, aircraft industry; Germany; Italy; Soviet Union
Evans, John 110, 149

F

F-4. *See* McDonnell Douglas
Failures. *See* Engine problems (aircraft); Engine problems (rocket)
Farley, John 58, 62
Farmer, Elmo 28
Federal Aviation Administration 99, 104
Fleming, William 76
Flight Propulsion Laboratory. *See* Glenn Research Center, former names
Fluorine. *See* Fuels
Four Burner Area. *See* Glenn Research Center, facilities
France,
    aircraft industry 99
Friedman, Harold 8

Fuels 53, 54, 55
    High-energy 62, 75, 76, 93
        Fluorine 62, 80
        Liquid hydrogen 62, 80, 81, 85, 87, 88, **89**, 91
            Danger 75, **76**, 81, 88, **89**
            Development 75, 82, **83**
            Project Bee 75
            See also Centaur Program; Pratt & Whitney, RL-10; Propulsion Systems Laboratory (PSL), components, fuel-handling system
        Liquid oxygen 75, 80, 82, 88, **89**, 91
        Methane 80
        Pentaborane 62
    Hydrogen peroxide 82
    JP-4 (jet fuel) 62, 82
    Liquid nitrogen 88, 91
    Solid 75, 92, 93, 94
    Isentropic (storable) 82, 90, 91
Full-Scale Engine Research (FSER) 125-127, 132

G

Garg, Sanjay 128
Garrett Corporation (Ryan Aeronautical Corporation) 109
    ATF3 **109**, 110, 124
    TFE 731 109, 110
Gelalles, Achille 16
General Dynamics 85
    F-16, **125**
    F-111 Aardvark (U.S. Air Force) **105**, 130
General Electric 3, 17, 67
    GE4 103
    J47 67, 129
    J73 vii, **10**, 67, **68**
    J79 **52**, **67**, 68
    J85 94, 100, 101, **103**, 105, 125, 130, 132
        J85-13 **39**, **94**, 103, **104**, 105, 106, 126
        J85-21 125, 126, **127**
    Schenectady plant 67
Germany 4, 5, 6
    Aeronautics and space technology 53
        V-1 and V-2. *See* Missiles
        See also Alsos Mission
    Test facilities 6
        Berlin-Aldershof 6
        Herbitus 99
        Peenemunde 54
        Rechlin 6
        Zeppelin 3, 4, 5
Gibson, James 63
Gilmore-Olson Construction 117

Glenn, John 7
Glenn Research Center 137, 138
    Accident Investigation Board 112
    Apprentice Program 56, 66
    Budgets and funding ix, 15, 132, 137
    Central air-handling system 9, 10, 42
    Construction 7, 17, **118**
    Divisions
        Airbreathing Engines Division viii, 80, 99, 100, 106, 124, 125, 137
        Chemical Energy Division, 80
        Chemical Rocket Division viii, 81, 85, 87, 91, 100
        Electric Propulsion and Environment Test Branch 56
        Engine Research Division viii, 17, 28, 55, 61, 65, 69
        Fabrication and Property Division 56, 82
        Facilities Division 145
        Fluid System Components Division 101, 106, 125
        Fuels and Lubrication Division 75
        History Office ix, xiii, 145
        Propulsion Systems Division 77, 91, 137
        Research Facilities Panel 14, 15
        Rocket and Aerodynamics Division 87
        Technical Services Division 17
        Test Installations Division viii, 28, 36, 37, 56, 91, 100, 101
        Wind Tunnels and Flight Division 101, 106
    Facilities ix
        Administration Building 13, 137
        Altitude Wind Tunnel (AWT) ix, **8**, 9, 10, **20**, 33, 61, 67, 92, 129, 138, 145, 147
        B-2 (at Plum Brook) 91
        Chemistry Laboratory 76
        Engine Components Research Laboratory 76
        Engine Propeller Research Building (Prop House) **2**, 7, 76
        Engine Research Building (ERB) 7, 9, **20**, 42, 48, 80
            Dynamotor rig 5, 7
        Fabrication shop 82
        Four Burner Area **9**, 10, 15, 61, 65, 117
        Hybrid Computer Laboratory 131
        Icing Research Tunnel ix
        Instrument Laboratory 76
        Machine shop 82
        Plum Brook Reactor Facility ix
        Power Systems Facility 80
        Propulsion Science Laboratory. *See* Propulsion Systems Laboratory (main entry)
        Propulsion Systems Laboratory. *See* Propulsion Systems Laboratory (main entry)
        Rocket Engine Test Facility ix, 75, 81
        Rocket Laboratory 75, 80, 137
        Space Power Facility 59
        Space research facilities 77
        Special Projects Laboratory 107, 108
        Zero Gravity Facility 100

Glenn Research Center cont.
- Facilities cont.
    - 8- by 6-Foot Supersonic Wind Tunnel 62, 69, **76**, 103, 128, 137
    - 10- by 10-Foot Supersonic Wind Tunnel 17, 103, 105, 137
- Former names
    - Aircraft Engine Research Laboratory (AERL) ix, 7-9, 53-54, **118**
    - Flight Propulsion Laboratory 7, 13
    - Lewis Flight Propulsion Laboratory vii, 7, 13, 15, 17, 33, 55, 75, 77
    - Lewis Research Center 7, 75, 77, 80, 94, 99, 103, 117, 137
- Inspections 13, 14, 32, **33**, **34**, **67**, 76, **94**, 149
- Reorganizations 28, 53, 75, 77, 91, 99, 137
- Staff
    - Apprentices 56, 66
    - Glenn Historic Preservation Officer ix, 145, 149
    - Management ix
    - Model builders 56
    - Photographers vii, ix, 137, 147
    - Pilots 127
    - *See also* Propulsion Systems Laboratory, staff
- Visitors 33, 76

Goddard Space Flight Center 17
Godman, Robert **100**
Goett, Harold 17
Goette, William 80
Gold, Harold **107**
Groesbeck, John 69
Ground Observation Corps vii, 137
Grumman
- F-14 Tomcat (U.S. Navy) 105, 131

H

Hall, Jesse **14**
Hallion, Richard 130, 149
Hannum, Ned **85**, 91
Hart, Clint 62
Hazardous materials remediation. *See* Propulsion Systems Laboratory, demolition, abatement
Henry Pratt Company 24
Herbitus. *See* Germany, test facilities
Highly Maneuverable Aircraft Technology (HiMAT). *See* Rockwell International, HiMAT
Himmel, Seymour **33**
Hispano-Suiza 6
Historical preservation ix, 140, 145, 147
- Documentation xiii, ix, 145, **147**, 151, 153
- Glenn Historic Preservation Officer. *See* Glenn Research Center, staff
- History Office. *See* Glenn Research Center, divisions
- National Historic Preservation Act 140, 145
- Ohio State Historic Preservation Officer ix, **145**
- Publications ix, x, 145, 147, 149, 153, 155-163

H.K. Ferguson Company **20**

Honeywell International 109, 130
Hong, William 132
Huber, John 109, 110
Hunsaker, Jerome 13
Hurrell, George 61, 62
Hybrid Computer Laboratory. *See* Glenn Research Center, facilities
Hydrogen. *See* Fuels

I

Icing Research Tunnel. *See* Glenn Research Center, facilities
Ide, John Jay 6
Inspections. *See* Glenn Research Center
International Geophysical Year vii
Iroquois engine. *See* Avro Aircraft Limited
Italy
    Fiat research facilities 3

J

Jacobs, Eastman 17
Jaw, Link 128
Jet engines 9, 10, 14, 44, 56, 67, 77, 99-112, 124, 129
    See also Engines; Engine components (aircraft); Engine problems (aircraft); Turbojet engines
Jet Propulsion Laboratory 75
Johnsen, Irving 91

K

Kemper, Carlton 14, **53**, 54
Kobak, John **74**, 75, **80**, 81, 82, **83**, **86**, **87**, 88, 91, 100, 149
Korean War vii, 13, 15, 56, 68
Krier, Gary **129**
Kutina, Frank 100, **101**

L

Langley Research Center (Langley Memorial Aeronautical Laboratory) 6, 53
Latto, William, Jr. 104
Lehtinen, Bruce 131
Lewis, George 6
Lewis Research Center. *See* Glenn Research Center, former names
Liberty engine. *See* National Bureau of Standards
Lindberg, Charles 6
Linear quadratic regulator. *See* Engine components (aircraft), controls
Lockheed (and Lockheed Martin)
    F-16 Falcon **125**
    F-104 Starfighter 68
    U-2 108
    X-16 Falcon 125
Looft, Fred 81, **100**
Low Cost Engine **107**, 108, 124
Lundin, Bruce 16, 28, 112, 117, 137

M

Main, Leslie 149
Mansour, Ali **85**, 91
Marquardt
    RJ43 ramjet **64, 65**
    Test facilities at Van Nuyes 64
Marshall Space Flight Center 88, 91
    Army Ballistic Missile Agency 85
Martin
    B-57B Canberra 75
McAulay, John 69, **70**
McCarthy, Eugene 137
McCook Field 3
McDonnell Douglas
    F-4 Phantom **68**
Meilander, Erwin "Bud" **28**, 66, 91, **102**
Meyer, Carl 61, 91
Missiles
    Defense 29, 53, 54, 64, 67, 76
    Interceptor 64. *See also* Navaho Missile Program; Boeing, Bomarc missile
    Intercontinental 61, 63, 69. *See also* Navaho Missile Program
    Minuteman 65
    V-1 and V-2 54
Mitigation. *See* Historical preservation
Morris, Joseph **145**
Muroc Lake. *See* Dryden Flight Research Center
Murray, Russ 54, 149

N

NACA. *See* National Advisory Committee for Aeronautics
NASA. *See* National Aeronautics and Space Administration
National Advisory Committee for Aeronautics (NACA) 3, 4
    Budget 13, 15
    Executive Committee 4, 6
    Powerplants Committee 4, 6
    European Office 6, 13
National Aeronautics and Space Agency (NASA)
    Budget ix, 94
    Established 77
    Foundation of 69, 77
    Headquarters 138
    History Publications ix, 7, 149, 155-163
    See also Ames Research Center; Dryden Flight Research Center; Glenn Research Center; Jet Propulsion Laboratory; Marshall Space Flight Center
National Aircraft Show in Dayton 68
National Bureau of Standards (NBS) **4**, 5, 6, 7
    Altitude Laboratory 5, 6
    Dynamotor Laboratory 5
    Liberty engine **5**, 6

National Register for Historic Places ix, 145
Nationalization of aeronautical research 4
Navaho Missile Program vii, 29, 54, **63**, 65, 69, 147
    Canceled 63
    Failures 63
    G-38 54, **55**, 63
    Navaho II (G-26) 54, 55, 63
    X-10 54, **55**, 63
Navy, U.S. 4, 65, 105, 107, 108, 131
    Aircraft. *See* Grumman
    Naval Weapons Center 108
NBS. *See* National Bureau of Standards
Neidengard, Bill **90, 92, 93**
Nitrogen. *See* Fuels
Noise reduction. *See* Engine problems (aircraft)
North American Aviation 54, 63
    F-15 Eagle **131**, 132
    F-86 Sabre fighter 67, 68
    F-86H Sabrejet vii, 67, 68
    X-15 Eagle 125
    See also Navaho Missile Program
Northrop
    F-5 Freedom Fighter 105
    T-38 Falcon 105
Norton Company 78
Nozzles. *See* Engine components (aircraft); Engine components (rocket)

O

Ohio Historic Preservation Office. *See* Historical preservation
Open House. *See* Glenn Research Center, inspections; visitors
Operation Desert Storm 125
Operation LUSTY. *See* Alsos mission
Operation Paperclip. *See* Alsos mission

P

Pearl Harbor vii
Pelouch, James 92
Pentaborane. *See* Fuels, high-energy
Perhala, George 32
Peters, Daniel 69
Pinkel, Benjamin 7, 8, 9, **14**, 15, 16
Pittsburgh-Des Moines Steel Company 117
Plum Brook Reactor Facility. *See* Glenn Research Center, facilities
Power Systems Facility. *See* Glenn Research Center, facilities
Pratt & Whitney ix, 87, 125, 130, 131
    F100 101, 125, **126**, 127, 131, 132, 147
    FY-401 and F-401 131
    RL-10 ix, **74**, 75, 80, 81, **85**, **86**, 87, 88, 91, **94**, 145, 147
        Development 85

Pratt & Whitney, cont.
- RL-10, cont.
  - PSL testing 86-88
 - Test facilities 87
 - TF30 **98, 100**, 101, 105, 125, 132
   - TF30-P-1 105, 106
   - TF30-P-2 105
   - TF30-P-3 106
   - TF30-P-9 130

Project Bee. *See* Fuels, high-energy, liquid hydrogen

Project Bumblebee. *See* Ramjets

Prop House. *See* Glenn Research Center, facilities, Engine Propeller Research Laboratory

Propulsion Science Laboratory. *See* Propulsion Systems Laboratory

Propulsion Systems Laboratory (PSL) **vii, viii, x**, 10, **15**, 33, **35**, 56, 99, **118, 123, 124**, 132, **141**, 145, **147**, 149
- Altitude simulation viii, 16, 34, 48
- Components
  - Combustion air system 10, 16, **20**, 33, **42, 43**, 110, 111, 117, **119**, 138, 140, **142**, 153
    - Air dryer **44**
    - Hypersonic temperatures 77-78
  - Compressors 16, **26**, 28, 34, **43**, 121, **122**
  - Control room 34, 40, **46, 47**, 78, 81, **83, 89**, 91, **122**, 140, 153
    - Quiet 47
    - Unpressurized (dangerous) 75, 81. *See also* operation and test procedures
  - Cooling system 16, **27**, 28, 34, **49, 50**
    - Cooling tower. *See* support buildings
    - Cooling vanes 140, **143**
    - Primary coolers **12, 19, 21, 22, 23, 29, 49**, 102, **120, 123, 136**, 140, **143**
    - PSL No. 3 and 4 cooling system 117, **119**, 120
    - Secondary cooler **50, 123**
  - Desiccant air dryer **123**
  - Electrical power 36, 41, 42, 117, 138. *See also* test schedule
  - Exhaust system viii, 10, 15, 16, 28, 34, **49**, 88, 153
    - Atmospheric air vent **139**
    - Common plenum (PSL No. 3 and 4) **121**
    - Exhausters **26, 48**, 112, 121, 122
  - Fuel-handling system 36, 78, **89**
  - Heating (air) **27**, 28, 34, 43, **44**
  - Instrumentation (thermocouples, pressure rakes, etc.) 34, **37, 39**, 45, **58, 59**, 93, 98, 101, **126**, 138
  - Pipes and valves **41, 43**, 81, 117, **120**, 137, 138
  - Refrigeration (air) 16, 34, **44**
  - Safety equipment 138
  - Test chambers 10, 23, **24**, 34, **35**, 40, **42**, 45, **62**, 65, 74, 79, **101**, 116, 117, **119, 120, 121, 123, 126, 127**, 139, **144**, 146
    - Cowl, bellmouth **45, 58**
    - Direct-connect mode **39**, 55, 57, 62
    - Free-jet mode **39**, 55, 57, **61**, 65
    - Flamespreader. *See* modifications
    - Lid (clamshell hatch) 24, **38**, 40, **42, 61**
    - Thrust stand 16, **39, 45**, 78, 87
- Construction vii, ix, **12**, 16, **19-28**, 117

Propulsion Systems Laboratory (PSL) cont.
    Construction cont.
        Budget 15, 16
        Design 16, 17, 117
        Phases 16
    Damage 88
    Demolition ix, **136**, 140, **141-144**, 145
        Abatement 140
        Decision 117, 138, **139**, **140**
        Plan 138, 140
    Model and drawings **15**, **32**, **33**, **34**, **35**, **40**, **45**, **119**, **123**
    Modifications
        Compressors and exhausters updated 117
        Conversion to office space 137, **138**
        Electronic units to replace manometers 122
        Flamespreader ix, **102**
        Gaseous hydrogen burner 106
        Large check valves 112
        Pebble bed heater ix, **78-79**
        Screen at diffuser outlet 58, 125, 128
        Vortex generator 58
    Need for and plans for ix, 10, 14, 33, 54, 67, 147
        Importance of full-scale testing 33-34, 37, 67, 76, 147
    Operation and test procedures 33, 34, **35**, 37-50, 75
        Dangerous 75, 81, 82, 88, **89**, 91
        Failures 82, **83**, 88
        Overnight and shift work. *See* test schedule
    Shutdown ix, 137
    Staff
        Aircraft model makers 56
        Coordination 34, 36
        Electricians 34, 36, **37**, **38**, 39, 40, 41, 48, 58, 65, 70, 77, 81, 86, 90, 91, 100
        Mechanics 28, 34, 36, **38**, **39**, **40**, 47, 55, 56, 57, **58**, 66, **90**, 91, 100, 103, 137
        Operations engineers 41, **42**, **48**, 74, 88, 89, 91, **101**, 110, 147
        Researchers viii, ix, 8, 9, 28, 34, 36, **37**, **38**, **40**, 41, **52**, **59**, 61, 62, 65, 66, 68, 69, **70**, 75, 77, 80, 82, **85**, 87, 88, **90**, 91, 92, 103, 105, 106, **107**, 125, 126, 127, 128, 131, 147
        Shift supervisor 28, 36, 66, 76
        Technicians viii, ix, 14, 29, 34, 36, **65**, **70**, 74, 75, 77, **82**, 86, 91, 125, **126**, 137, 147
        Test engineers 34, 36, **38**, 40, **46**, 47, 55, 57, 74, 75, 82, **86**, **91**, 100, **101**, 112
    Support buildings 153
        Circulating Water Pump House 28, **50**
        Cooling tower **27**, **50**
        Equipment Building **viii**, ix, **15**, **32**, **35**, **118**, 121, **122**, **123**, **124**, 137
            Equipment Building components 43, 44, 48, **50**, 117
            Construction 20, **21**, 23, **25**, **26**, **27**, 28
            Explosion 88, **110**, **111-112**
            Operation 36, 42
        Fuel Storage Building **123**
        Heater Building 117, 123
        Operations Building **37**, **76**, 100

Pursuit of Power

Propulsion Systems Laboratory (PSL) cont.
    Support buildings cont.
        Service Support Building **15**, **32**, **35**, **123**, 140, **142**
        Shop and Access Building vii-viii, **15**, 28, **32**, **35**, **40**, **43**, **45**, 47, 76, **89**, **123**, 137, **138**, 140, **142**, 143
            Construction **12**, **21**, **23**, **25**
            Control Room. *See* components
            Test area **38**, **45**, **47**, **90**, **94**
            Test chambers. *See* components
            Shop 37, **52**
    Test Programs
        Avro Iroquois PS.13. *See* Avro Aircraft Limited
        Garrett ATF3. *See* Garrett Corporation
        Garrett TFE21. *See* Garrett Corporation
        General Electric J73. *See* General Electric
        General Electric J79. *See* General Electric
        General Electric J85. *See* General Electric
        Hydrogen peroxide rocket. *See* Fuels
        Low Cost Engine. *See* Low Cost Engine
        Marquardt XRJ47 (Bomarc missile). *See* Marquardt
        Pratt & Whitney F100. *See* Pratt & Whitney
        Pratt & Whitney RL-10 rocket. *See* Pratt & Whitney
        Pratt & Whitney TF30. *See* Pratt & Whitney
        Storable propellant. *See* Fuels, solid, isentropic (storable)
        Wright Aeronautical XRJ47 (Navaho Missile). *See* Wright Aeronautical Division; Navaho Missile Program
        260-inch solid rocket. *See* Aerojet
    Test schedule (overnight testing, shift work, no time off) 36, 40, 41, 42, 66, 75, 81, 110, **132**, 137
Propulsion Systems Laboratory No. 3 and 4 **118**, **119-124**, 132, 137, 138, 147
    Construction ix, **116**, **119-122**
    Design 117, **119**
    Need for 117
    Operations ix

Q
Quiet Engine **107**

R
Ramjets **8,** 44, 53, 54, 55, 56, 57, 61, 62, 67, 70, 77
    Cooling and dissociation 78
    Design 54-55
    Dynamic response 62
    Fuel types 56
    Project Bumblebee 65
    Shock positioning 63, 65
    See also Wright Aeronautical Division, XRJ47; Marquardt
Rayle, Warren 58, 61
Rechlin. *See* Germany, test facilities
Reciprocating engines 55
Redstone booster 63

Reid, Henry 53
Reilly, Dwight 57, **59**
Reino, Salmi 92
Rennick, Paul 28, 66
Republic Steel 56, 81
Restarting at high altitude. *See* Engine problems (aircraft), flameout
Ricciardi, Nicholas 56
RL-10. *See* Pratt & Whitney
Robinson, Russell 53
Rocket Engine Test Facility. *See* Glenn Research Center, facilities
Rocket fuels. *See* Fuels
Rocket Laboratory. *See* Glenn Research Center, facilities
Rockets 53, 54, 67, 70, 75-77-79, 99
    Chemical rockets **82**-94
    Component testing 76
    Performance 91
    Solid rockets 91, 92, **93**, 94
    See also Atlas; Centaur; Pratt & Whitney, RL-10; Redstone; Saturn V; Thor; Titan
Rockwell International 127
    HiMAT (fighter technologies) 127, **128**
Roots-Connersville Corporation 16, 26, **48**
Ross, Al 91
Ross, Larry 88, 149
Ross Heater Company 16
Rothrock, Addison 53
Russia. *See* Soviet Union
Ryan Aeronautical Corporation. *See* Garrett Corporation

S

Saari, Martin 33, 34, **37**, **67**, 76
Saturn V (rocket) 91
Schey, Oscar **14**
Schroeder, Rudolph **3**
Schulke, Richard 32
Seashore, Ferris 61
Sharp, Edward "Ray" **13**, **14**, 16, 56, 77
Silverstein, Abe 53, 75
    Center Director 77, 88, 99
    Director of Research **13**, **14**, 17
    Headquarters 17
Skywatchers. *See* Ground Observation Corps
Smith, Ivan 58, 61
Solar power 80
Soviet Union vii, 29, 53
    Aeronautics 4, 64, 99
    Space program 76
    See also Cold War
Space Station Freedom 80
Special Projects Laboratory. *See* Glenn Research Center, facilities

Sputnik I 63, 76
State History Preservation Officers. *See* Historical preservation
Steele, Jerry 109, 110, 149
Stirling engine 80
Stratton, William **4**, 5, 6
Supersonic Transport Technology (SST). *See* Boeing
Surveyor 85, 91
Systems Control, Inc. 131

T

Teledyne Ryan Technologies Inc. 111
    YQM-98 A Compass Cope Tern **108**, 109, 110
Test chambers. *See* Altitude research; Propulsion Systems Laboratory, components
Thomas, Albert 13, 14, 16, 29, 76
Thomas, Wayne **91**
Thor (booster) 63
Titan (rocket) 90, 91
Treadwell Construction 16
Trout, Arthur 91
Turbofan engines ix, 98, 99, 105, 109, 111, 129, 132
    Design 99, 109-110, 129, 132
    See also Pratt & Whitney, TF30; Garrett Corporation
Turbojet engines 99, 103, 105, 128, 129, 145, 147
    Axial-flow 53, 54, 67, 124
    Centrifugal 53
    Development ix, 8, 9, 55, **67**, 68, 70, 127-128, 132
Turbosupercharger. *See* Engine components (aircraft)

U

Unitary Wind Tunnel Plan 13, 17
University of Michigan Aeronautical Research Center 64
U.S. Space and Missile Museum 63
USS Nautilus 63

V

Van Allen, Harvey vii
Variable-area exhaust nozzle. *See* Engine components (rocket)
Vasu, George 62
Victory, John 76
Vietnam War 68
von Braun, Wernher 54, 88
Vought
    F-8 Crusader **129**

W

Wagoner, Lee **104**
Walker, Robert 10, 28, **46**, 48, 54, **55**, **56**, 57, **61**, 67, 117, 149
Wanhainen, John 91
Wasielewski, Eugene 16, **17**, 28, **63**, 64
Weapon System 104A. *See* Navajo Missile Program

Welna, Henry 57-62
Wentworth, Carl 58, 61, 65
Wenzel, Leon **106**
Werner, Roger 126
Westinghouse 55, 67
    J40 54, 63
White, Jack 109
Williams, Dan 16
Wind tunnels 63-64
    Cost 61
    Design viii, 9
    Hypersonic (in PSL) 78
    Supersonic 103, 105
        8- by 6-Foot Supersonic Wind Tunnel. *See* Glenn Research Center, facilities
        10- by 10-Foot Supersonic Wind Tunnel. *See* Glenn Research Center, facilities
    See also Unitary Wind Tunnel Plan
Wine, Howard 10, 36, 41, **46**, **66**, 76, 100, **146**, 149
Wingenfeld, Neal 36, 47, 75, **81**, 85, 91, 100, 137, 149
World War I 4, 5, 15
World War II 14, 15, 53, 54, 55, 67
Wright Aeronautical Division
    J65 55
    Wood-Ridge facility 55, 64
    XRJ47 29, 54, 55, 57-58, **59**, 60, **61**, 62-64
        Damage to liner 57
        Development 54-55
        Flameholders. *See* Engine components (aircraft)
        Ignition. *See* Engine problems (aircraft)
        PSL testing 57-62
        1-A classification 55
        48-inch ramjet **58**, **62**
    See also Curtiss-Wright; Navajo Missile Program
Wright Park 117, **118**

X

XRJ47. *See* Wright Aeronautical Division

Y

Yeager, Charles 15
Young, Alfred 28

Z

Zeppelin Aircraft Works. *See* Germany, test facilities
Zero Gravity Facility. *See* Glenn Research Center, facilities

www.ingramcontent.com/pod-product-compliance
Lightning Source LLC
Chambersburg PA
CBHW080543170426
43195CB00016B/2658